The Transnational Unconscious

The Palgrave Macmillan Transnational History Series

Series Editors: **Akira Iriye**, Professor of History at Harvard University, and **Rana Mitter**, University Lecturer in Modern History and Chinese Politics at the University of Oxford

This distinguished series seeks to: develop scholarship on the transnational connections of societies and peoples in the nineteenth and twentieth centuries; provide a forum in which work on transnational history from different periods, subjects, and regions of the world can be brought together in fruitful connection; and explore the theoretical and methodological links between transnational and other related approaches such as comparative history and world history.

Editorial Board: **Thomas Bender**, University Professor of the Humanities, Professor of History, and Director of the International Center for Advanced Studies, New York University; **Jane Carruthers**, Professor of History, University of South Africa; **Mariano Plotkin**, Professor, Universidad Nacional de Tres de Febrero, Buenos Aires, and member of the National Council of Scientific and Technological Research, Argentina; **Pierre-Yves Saunier**, Researcher at the Centre National de la Recherche Scientifique, France and Visiting Professor at the University of Montreal; **Ian Tyrrell**, Professor of History, University of New South Wales

Titles include:

Gregor Benton and Edmund Terence Gomez
THE CHINESE IN BRITAIN, 1800–PRESENT
Economy, Transnationalism and Identity

Glenda Sluga
THE NATION, PSYCHOLOGY, AND INTERNATIONAL POLITICS, 1870–1919

Joy Damousi, Mariano Ben Plotkin (*editors*)
THE TRANSNATIONAL UNCONSCIOUS
Essays in the History of Psychoanalysis and Transnationalism

Forthcoming titles:

Sebastian Conrad and Dominic Sachsenmaier (*editors*)
COMPETING VISIONS OF WORLD ORDER
Global Moments and Movements, 1880s–1930s

Matthias Middell, Michael Geyer, and Michel Espagne
EUROPEAN HISTORY IN AN INTERCONNECTED WORLD

The Palgrave Macmillan Transnational History Series
**Series Standing Order ISBN 978–0–230–50746–3 Hardback
978–0–230–50747–0 Paperback**
(*outside North America only*)

You can receive future titles in this series as they are published by placing a standing order. Please contact your bookseller or, in case of difficulty, write to us at the address below with your name and address, the title of the series and the ISBN quoted above.

Customer Services Department, Macmillan Distribution Ltd, Houndmills, Basingstoke, Hampshire RG21 6XS, England

---

The Transnational Unconscious

Essays in the History of Psychoanalysis and Transnationalism

Edited By

Joy Damousi
Professor of History, School of Historical Studies, University of Melbourne

and

Mariano Ben Plotkin
CONICET/Professor of History, Universidad Nacional de Tres de Febrero

Contents

Section 4   Challenging Centre and Periphery

Notes on Contributors

Alejandro Dagfal is professor at the University of La Plata and researcher at the CONICET, the national research council of Argentina. He has written extensively on the history of Argentine psychology and psychoanalysis during the twentieth century, comparing the reception of Anglo-Saxon and French traditions. His works include "La naissance d'une 'conduite à la française': de Ribot à Janet", *L'Évolution psychiatrique*, *67*(3), 591–600, Elsevier, Paris, 2002 and "La naissance d'une psychologie clinique 'd'inspiration psychanalytique'," *Psychologie Clinique*, *17*, 83–102, L'Harmatan, Paris, 2004.

Joy Damousi is Professor of History in the School of Historical Studies at the University of Melbourne. Her recent areas of publication are on the topics of memory and war, the history of emotions and psychoanalysis, themes she has explored in *The Labour of Loss: Mourning, Memory and Wartime Bereavement in Australia* (Cambridge, 1999); *Living with the Aftermath: Trauma, Nostalgia and Grief in Post-war Australia* (Cambridge, 2001); *History on the Couch: Essays in history and psychoanalysis* (co-edited with Robert Reynolds, Melbourne University Press, 2003), and *Freud in the Antipodes, a cultural history of psychoanalysis in Australia* (University of New South Wales Press, 2005).

Elizabeth Ann Danto is associate professor and chair of the Foundations of Practice Sequence at Hunter College School of Social Work – City University of New York. She is the author of *Freud's Free Clinics – Psychoanalysis & Social Justice, 1918–1938* (Columbia University Press, 2005) as well as numerous articles on the social history of psychoanalysis, and social welfare and employment policy.

Federico Finchelstein is a graduate student in History at Cornell University. He is the author of *Fascismo, Liturgia e Imaginario* (Buenos Aires: Fondo de Cultura Economica, 2002) and the editor of *Los alemanes, el Holocausto y la culpa colectiva* (Buenos Aires, Eudeba, 1999). He is now working on the relationship between Fascism and the reception of psychoanalysis in Argentina and Italy.

Frances Gouda is Professor of History and Gender Studies in the Department of Political Science of the University of Amsterdam in the Netherlands. She is the author of three books: *Poverty and Political Culture: The Rhetoric of Social Welfare in France and the Netherlands, 1815–1854* (1994), *Dutch Culture Overseas: Colonial Practice in the Netherlands-Indies, 1900–1942* (1995), and *American Visions of the Netherlands East Indies/Indonesia. US Foreign Policy and Indonesian Nationalism, 1920–1949* (2002). She is co-editor of *Domesticating the Empire: Race, Gender and Family Life in French and Dutch Colonialism* (1999).

Mariano Ben Plotkin is member of the National Council of Scientific and Technological Research (Consejo Nacional de Investigaciones Científicas y Técnicas, Argentina) and researcher at the Instituto de Desarrollo Económico y Social. He is also a professor at the Universidad Nacional de Tres de Febrero (Buenos Aires). His books include *Mañana es San Perón* (1994; English edition by Scholarly Resources, 2003) and *Freud in the Pampas* (Stanford 2001; Spanish edition 2003). He also edited several volumes on topics related to the history of psychoanalysis and of social sciences.

Jane Russo has a PhD in Social Anthropology from the Federal University of Rio de Janeiro and works as a professor and researcher at the Institute of Social Medicine of the State University of Rio de Janeiro (UERJ). She has written several articles on the history of the "psy" professions in Brasil, and is the author of two books on the subject "O corpo contra a palavra – o movimento das terapias corporais no campo psicológico dos anos 80" and "O mundo psi no Brasil". She has also co-organized the books "Duzentos anos de psiquiatria" (with João Ferreira da Silva Filho) and "Psicologização no Brasil: atores e autores" (with Luiz Fernando Duarte and Ana Teresa Venancio).

Sergio Eduardo Visacovsky is Professor of Anthropology at University of Buenos Aires, and Institute for Social and Economic Development (IDES) (Argentina), and a researcher at National Council for Scientific and Technical Research of Argentina (CONICET). He has studied how experiences of political process become social memory, contributing to construction of intellectual, academic and professional fields. His most recent books are: *El Lanús. Memoria y política en la construcción de una tradición psiquiátrica y psicoanalítica en la Argentina* (Buenos Aires, Alianza, 2002), and *Historias y estilos de trabajo de campo de campo en la Argentina* (Buenos Aires, Antropofagia, 2002), edited in collaboration with Rosana Guber. At present, he is researching on crisis experiences of urban middle classes.

Eli Zaretsky is Professor of History in the Graduate Faculty at the New School University in New York City. His book, *Capitalism, the Family and Personal Life* has been translated into 14 languages, and his articles on the history of the family, psychoanalysis, and modern cultural history have appeared in numerous scholarly journals. His latest book, *Secrets of the Soul: A Social and Cultural History of Psychoanalysis* was published by Knopf in 2005.

Acknowledgements

We would like to thank Dr. Darwin Stapleton and the wonderful staff of the Rockefeller Archive Centre, and especially Norine Hotchman, for their support and for fostering a stimulating environment of intellectual exchange. Thanks are also due to Warwick Anderson and Richard Keller, the conveyors of the 'Unconscious Dominions: Comparing Histories of Psychoanalysis, Empire, and Citizenship' workshop held at the University of Wisconsin-Madison in 2005, where we had the opportunity of meeting and discussing the possibility of this project the first time. Michael Strang and Ruth Ireland from Palgrave Macmillan have been very helpful and supportive in bringing this project together. Finally, without the energy and intellectual engagement of our contributors, this volume would not have been possible. We thank them all for their lively scholarship and providing us with this opportunity of a genuine transnational exchange.

Joy Damousi
Mariano Ben Plotkin

Foreword

"The Transnational Unconscious" is an intriguing title for a collection of essays on the history of psychoanalysis. How can the unconscious be transnational? On the surface it might be thought that human consciousness is a deeply personal phenomenon and cannot, therefore, be duplicated between individuals, let alone transferred across national borders. On reflection, however, it becomes clear that the very essence of unconsciousness is something that transcends national and other political divisions. It does not matter in which country one lives when speaking of one's unconscious desires, dreams, and inhibitions. There may be such a thing as national consciousness, but not national *un*consciousness. Since psychoanalysis deals with the unconscious, it must by definition be considered a transnational scholarly discipline. To be sure, all fields of scholarship are supposed to be universally verifiable and to know no national boundaries, but the history of such disciplines as economics, political science, sociology, and many others (including, particularly, history) shows that at least in their inception they were all embedded in national frameworks. Psychoanalysis was clearly different.

It is not surprising, therefore, and as several chapters in this book point out, that psychoanalysis, which first developed as a specialized field of science in Vienna at the turn of the last century, should have crossed nations and regions without difficulty and found itself replicated in many parts of the world. The focus in this volume is on the diffusion of psychoanalysis from Europe to Latin America, although some chapters deal with Australia, the Dutch East Indies, and the United States as well. The chapters show how a number of prominent European psychoanalysts were instrumental in founding the discipline in Argentina, Brazil, and other countries, and they examine why this particular discipline has flourished so successfully in South America. Transnationalization is not the same thing as universalization, and the authors explore how local academic and social conditions affected the ways in which psychoanalysis was put into practice in these countries. Indigenization and growth added further to the transnational character of psychoanalysis, showing that it could be adapted to local conditions; in other words, it lent itself admirably to transnationalization.

This leads to an interesting perspective on the history of psychoanalysis. It could be, and was indeed, a liberalizing force politically in the sense that it could free individuals from social and political contexts and constraints. As a study of the unconscious, it speaks to the human side of an individual's life, not to his or her political identity or social status. Such an inherently apolitical position does of course have political significance, particularly in authoritarian countries that demand individual subordination to the state or in hierarchically structured societies that differentiate people on the basis of race, gender, and other distinctions. Sigmund Freud was an ardent advocate of transnational civil society, while in Harlem in the US Richard Wright used Freudian theory as a theory of action for racial equality. In the Dutch colony of Indonesia (as Frances Gouda's chapter here shows), psychoanalysis had the potential to foster anti-imperial liberation, although it also provided justification for perpetuating colonial control by positing a primitive unconscious (presumed to be distinct from the unconsciousness of Europeans) that made the indigenous people unfit for self-governance.

All in all, the chapters in this volume make exciting reading for anyone curious about the transnationalization of scholarship or about the local political uses to which an inherently transnational discipline may be put. By providing specific examples of the ways in which psychoanalysis crossed the oceans, adapted to local cultures, and coped with volatile political conditions, the authors in this volume make a valuable contribution to the study of transnational history.

Akira Iriye
Rana Mitter

Introduction

Joy Damousi and Mariano Ben Plotkin

Dr. Pieter Mattheus van Wufften Palthe was a Dutch psychiatrist who characterized the post-World War II Javanese nationalist movement as a pathological manifestation of the Javanese's psyche. Richard Wright was an American black leftist intellectual trying to make sense of (and resist) the place of black people in American society by giving a voice to their subjectivity. What do these two stories have in common? Both Dr. Pieter Mattheus van Wufften Palthe and Richard Wright enlisted psychoanalysis as an analytic – and we could say political – tool to achieve their ends. While the former made an attempt to "psychoanalyze" and therefore delegitimize Indonesians' fight for independence, the latter, however, used psychoanalysis for the opposite purpose: as an instrument to reconceptualize the subordinate position of blacks and later of other colonial groups. In other words, van Wulfften Palthe used his understanding of psychoanalysis to rationalize his country's loss of its colonies; in contrast, Wright appropriated psychoanalysis as a liberating paradigm for the US black population and other oppressed groups. Which are the qualities of psychoanalysis as a body of knowledge and of its worldwide diffusion, that make it fit for such different appropriations?

It would not be an exaggeration to define the twentieth century as the "psychoanalytic century." If we consider the publication of Freud's foundational work *The Interpretation of Dreams* in 1900 as the birth of psychoanalysis,[1] then it is clear that the history of psychoanalysis was a successful one. This theoretical body – created in Vienna (a declining capital of a declining empire) by a Bohemian-born Austrian Jewish medical doctor who occupied a relatively marginal position in the academic and professional fields of his country, with the double purpose of curing certain mental disorders and at the same time researching the mind – in less than two decades became a transnational discipline,

1

anchored on a strong international institutional apparatus, that vastly expanded its original fields of application. In his *Autobiographical Study* (1925), Freud acknowledged the broad application of psychoanalysis to a large number of disciplines and fields, including the arts. By then, psychoanalysis had already transcended national and cultural boundaries and was practiced and discussed in countries as culturally and geographically removed from the new republic of Austria as Peru and India.[2] Although it is true, as historian Carl Schorske has argued,[3] that psychoanalysis is a child of its time and place, the fact is that it soon became a transnational system of beliefs and thought, expanding its parameters beyond national boundaries. During the last century psychoanalytic ideas have been historically influential in various forums for understanding interior life, the workings of the modern self, and accounting for human action. It has contributed significantly to understanding sexualities and gender identity and it has been argued that psychoanalysis provides crucial elements for the construction of modern subjectivity. Psychoanalysis has had an impact on a range of intellectual disciplines such as philosophy, psychology and anthropology and has informed a range of cultural endeavors in the creative arts such as film and literature. Throughout the twentieth century several avant-garde artistic movements fed on ideas originating in Freudian thought. During the last one hundred years it has provided a conceptual framework and a methodology that has been used to filter, interpret and construct different dimensions of reality. As John Forrester points out, the presence of psychoanalysis in the West (and not only there) is "so constant and pervasive that escaping its influence is out of the question." According to him, going back to pre-Freudian beliefs today, in spite of the existence of strong "anti-psychoanalytic" sentiment in many countries, including the US, is as likely as "going back to pre-Copernican beliefs" about the universe.[4]

The "internal" historiography of psychoanalysis that originated within the psychoanalytic community with Freud's own essay *On the History of the Psychoanalytic Movement* (1914), put an emphasis on the supposed "resistances" that, by definition and acting in the same way as patients do, society and culture opposes psychoanalytic discoveries. Despite this, it remains the case that very few systems of thought have been as influential or enduring as psychoanalysis has been in being disseminated throughout different cultures. If there is anything that requires an explanation it is the fast transnational success of psychoanalysis and its multiple appropriations, rather than the resistances it (as any other new system of thought) could have generated among some conservative sectors of the societies it entered. It could be said that transnationalism is one of the defining characteristics of psychoanalysis. The historiography on psychoanalysis (which has emerged as a field in itself) had traditionally focused on Freud and his immediate disciples. Unlike other disciplines, it seemed for a long time that the history of psychoanalysis could not be distinguished from the biography of its creator. Freud himself wrote two essays on the history of psychoanalysis: the already mentioned *On the History of the Psychoanalytic Movement* in 1914 and *Autobiographical Study* 11 years later, and in both, his own life and the history of the discipline he founded are virtually indistinguishable from each other.[5] This particular vision of the history of psychoanalysis was later taken by Ernest Jones in his monumental *The Life and Work of Sigmund Freud*, which laid the foundations for an "official history" of psychoanalysis.[6] Moreover, for Freud and his followers, the history of psychoanalysis is the history of the discipline's (and its practitioners') isolation and of the struggles against resistance. As could be expected, the hagiographic historiography of Freudians has provoked in more recent times an "anti-Freudian" reaction (linked to broader cultural developments) in which Freud is also at the center of the analysis but where his theories and practice are harshly criticized.[7]

More recently, a group of historians who could be characterized as "contextualists" have been focusing their analysis on the cultural, political and social conditions that made possible the appearance of psychoanalysis in late-nineteenth century Vienna. According to this version, psychoanalysis was not the brainchild of a single isolated genius but rather was the result of complex cultural, political and social factors operating in a particular historical setting.[8] The focus on psychoanalysis's Viennese origins, however, sheds light only on half of the story. The other half needs to be traced outwards as psychoanalysis was broadly disseminated through different countries and cultures. Only in the last few decades have new studies emerged which focus on the development of national psychoanalytic cultures and the reception and implantation of psychoanalysis in different cultural spaces.[9]

In this volume we focus on what is still a less studied dimension of the history of psychoanalysis: its international circulation and appropriation.[10] Moreover, we concentrate on geographical areas that are usually considered "marginal" and therefore left outside of the established historiography. If historicizing psychoanalysis is a difficult task because we are at the same time the students of the history of psychoanalysis and (in anthropological terms) the "natives," that is to say the object of our study – as intellectuals educated in a "psychoanalytic culture" – analyzing its transnational dimension poses additional methodological challenges.

At this point we should define our terms. While psychoanalysis is over a century old, transnationalism as a theoretical concept has a very recent past.[11] "Psychoanalysis" itself is not devoid of ambiguity. Psychoanalysis was born as a psychological theory and as a therapeutic technique, but throughout its life it has become a cultural artifact in the broader sense. Since we believe that the history of a system of ideas can not be distinguished from the history of its multiple receptions and appropriations, we refuse in this volume to accept the existence of an orthodox version of psychoanalysis that could be used as a yardstick to define deviations and heterodoxies. We are more interested in psychoanalysis as a widely defined cultural phenomenon than as a specific psychological theory and therefore we define as psychoanalysis all discourses and practices that are legitimized in their reference to a Freudian heritage. We think that our object will gain in richness what is looses in specificity. Furthermore, even if we took a more restricted definition of psychoanalysis things wouldn't be easier. Until the 1960s the International Psychoanalytical Association (IPA) held the monopoly over the legitimate practice of psychoanalysis. Since then, however, a very active alternative movement of followers of Jacques Lacan has emerged in France and in Latin America that dispute the IPAs hegemony and even its vision of what psychoanalysis is and should be.[12] Furthermore, the "Lacanian" movement itself is crossed by tensions and ruptures.[13]

The concept of "transnational" addresses both a quality of an object of study and a particular historical approach which focuses on movements, flows, circulation and intersection of people, ideas and goods across political and cultural borders. As Isabel Hofmeyrs points out, a transnational approach emphasizes the ways in which "historical processes are constructed in the movement between places, sites and regions."[14] In the context of this volume we would like to propose that a system of thought becomes transnational if it fulfills at least the three following criteria: first, if it circulates across national and cultural boundaries; second, if its analytic units transcend cultural limits; and third, if the center of production and diffusion (and the languages in which it is disseminated) change over time and therefore its development is not attached to any particular national space. Judged by this criteria, psychoanalysis clearly qualifies as a transnational system of thought. Although throughout its history psychoanalysis has been characterized – mostly by its detractors – as a German or a Jewish discipline, the fact is that it has transcended any particular cultural space. Its ideas (and practitioners) have circulated throughout the world, sometimes organizing psychoanalytic communities in remote locations, at times under circumstances not of their own choosing.[15] It could be said that immigration (willing or unwilling) has become one of the defining features of psychoanalysis. In countries like the US, the UK, Canada, Australia and to a lesser extent Argentina and Brazil, forced European immigration in the 1930s and 1940s played a crucial role in the constitution of the local psychoanalytic movements. In their contributions to this volume Joy Damousi, Alejandro Dagfal and Sergio Visacovsky, for instance, focus on three cases of transnational circulation. Damousi concentrates on the Hungarian analyst Andrew Peto who not only practiced psychoanalysis, but actively promoted it in Australia and in the US after leaving his native Europe; Dagfal writes on the diffusion of psychoanalyst Melanie Klein's theories in Argentina and France (the latter via Argentina); and Visacovsky studies how Argentines and Spaniards constructed different genealogies to legitimize the introduction of Jacques Lacan's version of psychoanalysis despite the fact that the reception of his theories in both countries were the result of the work of the same Argentine analyst.

Moreover, psychoanalysts claim that psychoanalytic categories such as the unconscious, sexuality, the Oedipus complex and others transcend cultural boundaries and are therefore universal. While these claims have become controversial, from the psychoanalytic point of view, its categories are not inherently attached to, or defined by, any particular cultural setting. Frances Gouda's chapter on the colonial application of psychoanalytic categories to the Javanese pro-independence movement and what we know about the reception of psychoanalysis in India, Japan and China show that the application of psychoanalytic categories are not indeed restricted to the Western world.

Finally, since its origins in German speaking *fin-de-siècle* Vienna, the centers of production and consumption of psychoanalysis shifted after World War II first to the English-speaking word, particularly Great Britain but, above all, the US and, more recently, to France and especially to Latin America. In this process psychoanalysis adapted to the different cultural settings in which it was implanted and, at the same time, had an influence on the receiving cultures and societies. In the last decades Argentina (particularly the city of Buenos Aires) has become recognized as the "world capital of psychoanalysis." Counting the IPA analysts and Lacanians of different persuasions, Buenos Aires houses one of the largest analytic communities in the world. Moreover, psychoanalysis has had a deep impact in the culture of the city.[16] In the 1940s, when the Argentine Psychoanalytic Association was created, Freud disciple (and then president of the IPA) Ernest Jones advised its founding members that the knowledge of the German language, while still desirable for a psychoanalyst, was losing its relevance as English was becoming the "official" tongue of psychoanalysis. Today, most psychoanalysis in the world (particularly that of Lacanian persuasion) is practiced and consumed in Spanish, French and, to a lesser extent, Portuguese. Although it is true that English is still the IPA's lingua franca, for the large Lacanian community, the official languages are French and Spanish.[17] As the editor of the recently published *Livre noir de la psychanalyse* points out, Argentina and France are today the two most "psychoanalyzed" countries in the world.[18]

The chapters of this volume focus, from a historical perspective, on the transnational dimension of psychoanalysis through a range of different contexts. Of course it is not our purpose to offer in a single volume a comprehensive history of psychoanalysis from a transnational perspective. In this book we have chosen case studies that focus on specific problems associated with a transnational approach to the history of psychoanalysis. Two questions in particular, however, lie implicitly behind the volume as a whole: one is theoretical and the other methodological. The first is: why have certain intellectual currents or systems of ideas become transnational while others fall into oblivion? At some points in history the intellectual movements of eugenics, spiritualism, theosophy, or even some contemporary theories of the mind like Pierre Janet's have been more influential than psychoanalysis, yet they did not retain their status. The second question, the methodological one is: how does one examine the transnational dimension of such an enduring system? What methodological frameworks can be used to discuss circulation, fluidity, exchange and hybridity? Transnationalism encourages a move away from traditional analytic paradigms to those which are framed by intersection and interdisciplinarity, challenging accepted categories such as center and periphery.[19] Transnational history aims to move between and across categories of analysis such as culture, power, identity, citizenship and nation, rather than remain fixed within the boundaries that define them.[20]

Unlike modern physics, biology or even medicine, the status of psychoanalysis as a science has been (and now it is more than ever) widely contested. Although Freud himself was adamant to establish psychoanalysis as a scientific and coherent theory, psychoanalysis itself is a moving category between academic and intellectual fields of analysis. In France and in Latin America, Lacanian psychoanalysts have moved away from medicine and from psychology. For them psychoanalysis belongs to the realm of the humanities. Many Lacanian analysts have a training in such disciplines as literature or philosophy (Miller himself is among the latter) or no formal academic training at all. It could be argued that its plasticity has been one of the factors that explain its diffusion. However, it fits comfortably neither among the natural sciences nor among the social sciences, although it has informed the latter. In fact, it could be said that psychoanalysis's "discoveries" and assumptions are a collection of unproven (and probably impossible to prove) hypotheses. It is, therefore, not its belonging to the realm of science that justifies psychoanalysis's transnational success. Moreover, as a therapy, psychoanalysis could never establish its higher rate of efficacy vis-à-vis other similar (or not so similar) therapeutic techniques. What is then, if not its scientific status or its clinical efficacy that explains its unprecedented level of influence, across time, place and culture?

As Peter Berger, Eli Zaretsky and others have shown,[21] psychoanalysis as a body of knowledge fits very well into what could be conceptualized as the "second wave of modernity," that is to say, modernity linked to industrialization, mass production and consumption ("Fordism," in Zaretsky's words), the decline of the family as a locus of production and the concomitant emergence of a new subjectivity.[22] In particular, Zaretsky points out that psychoanalysis built its theory precisely in the tension between supposedly universal patterns of human life such as the Oedipus complex, the unconscious and sexuality, on the one hand and the particular way in which the "new individual" originating in this "second wave of modernity" process these patterns on the other hand.[23] Under this view, psychoanalysis was the child of the "second modernity" and was, at the same time what provided its conditions of possibility. Something similar, of course, could be said about the origin of "modern social sciences" like sociology or economics.[24] This kind of general explanation, however, leaves aside the fact that by the turn of the twentieth century there were other psychological systems available in the "market of ideas" that could have played a similar role (including some of those proposed by Freud's former followers such as Alfred Adler and Carl Jung) and yet they never became transnational in the sense defined above. Not even those practicing in the social sciences could claim to be that successful. Emile Durkheim, for instance, although very influential during his lifetime, was not considered as a "classic" author until decades after his death. His poor reputation among American sociologists, for instance, was not reversed until Talcott Parsons's approach to sociology revived the interest in his work. Something similar could be said about the works and ideas of Max Weber who during his lifetime was not considered by his contemporaries as a sociologist but rather as a historian, and as an economist. The construction of a sociological canon is a post-World War II phenomenon.

A different conceptual explanation for the dissemination of psycho-analysis has been provided by Sherry Turkle in a pioneering book on the development of Lacan's psychoanalysis in France. Turkle focused on the emergence of what she calls a psychoanalytic culture defined as "the way psychoanalytic metaphors and ways of thinking enter every-day life."[25] Turkle calls "appropriability" the quality that a certain theory (in this case psychoanalysis) has, that facilitates the emergence of a cul-ture. According to Turkle, psychoanalysis as well as other sciences of the mind offer almost tangible "objects to think with": slips, dream anal-ysis and so on. Moreover, psychoanalysis has an inherent malleability that turns it into a flexible body of knowledge. The plasticity of psycho-analysis is placed in evidence by the existence of different "national" versions of it, many of them incompatible with each other and, at the same time, all of them claiming their legitimacy in their belonging to a Freudian genealogy (real or imaginary). In Turkle's view, nevertheless, the growth of a psychoanalytic culture is linked to three factors: first to its intrinsic qualities and malleability: psychoanalysis is a system that can give origin to a culture; second to the presence of what Philip Reiff has called "a moment of social deconversion," that is to say a time of rapid mobility and social dislocation; and third to its ability to provide a way for people to "think through" issues of political and social iden-tity.[26] To this we could add the existence of a devoted body of "diffusors" or "apostles" (meaning in this case people or institutions) that dissem-inate psychoanalytic thought through different means and in different cultural settings.

What is clear is that the transnational aspect of a system of thought cannot be properly grasped if at the same time one does not focus on different historical cases of reception and circulation of those ideas. The transnational and the local are deeply intertwined. Otherwise, how can we understand the emergence and development of such different versions of psychoanalysis as distinctively French or American? The transnationalization of a system of thought can only be understood as a historical process that occurs as a result of movements between sites, places and cultures but that, at the same time, is linked to local conditions. As Ulf Hannerz points out, flows of people and ideas have directions, and, we could add, the nature of those directions is the result of both global and local conditions.[27] Works that focus on the first aspect tend to ignore the second one and vice-versa. The chapters in this volume approach the transnationalization of psychoanalysis from this dual perspective, focusing on particular moments, people, episodes and processes that enlighten the local–global tension in the development of transnational psychoanalysis. Joy Damousi, for instance, concentrates on the activities of the Hungarian analyst Andrew Peto, himself a "transnational diffuser of psychoanalysis," who was active in his native Hungary, Australia and the US. A consideration of Peto's international movements highlights the paradox of the complex ways in which psychoanalysis can work at both the level of a national and transnational theory and practice. These dynamics are inter-related. For the traveling psychoanalyst, an examination of issues such as juvenile delinquency, child's play, countertransference and dream analysis – to take a few topics of Peto's writings – transcended national borders but at the same time were shaped in a specific national context. As a doctor, Peto was trained to apply psychoanalysis clinically and scientifically, and as a set of universal principles, therefore it was effortless for him to move professionally between Budapest, Sydney and New York. Peto resisted any political interpretation of psychoanalysis, despite the fact that his clinical experience was shaped by the traumas of war. Mariano Plotkin, on the other hand, looks comparatively at the early patterns of reception and circulation of psychoanalysis in two Latin American countries: Argentina and Brazil, before the establishment of the IPA-affiliated associations in each country, that is to say before an orthodoxy was established and therefore before the "psychoanalytic boom" of the 1960s and 1970s. He shows how the particular manner in which psychoanalytic thought was understood and appropriated in each country by medical doctors, artists and social scientists was linked to particular conceptions of "modernity." Is was also linked to historical social concerns and issues of national and racial identity, the internalization of which constitutes components of what sociologist Norbert Elias has characterized as the "national habitus."[28] Unlike their Argentine counterparts who believed that Argentina (and particularly the city of Buenos Aires) had, by the first decades of the twentieth century, crossed the threshold of Western modernity, Brazilian intellectuals, obsessed with the perceived "racial inferiority" of their country, looked to psychoanalysis for a way out to racial determinism.

As it was suggested at the beginning of the introduction, the situation in the Netherlands was very different. There, as Frances Gouda points out, psychoanalytic ideas were appropriated by members of the psychiatric colonial establishment to justify colonialism and to de-legitimize movements for independence in the colonies. Thus, for some Dutch psychiatrists, psychoanalysis became an important component of the transnational movement of ideas used to rationalize colonialism. Javanese independentists were characterized as immature neurotics. Gouda's chapter also implicitly brings about an important discussion regarding psychoanalysis: its supposedly intrinsic liberating character. Many scholars, particularly those coming from within the psychoanalytic community, have argued that psychoanalysis has a constitutive subversive and liberating nature and therefore can only implant itself in conditions of political and social freedom. Other scholars have emphasized the strong anti-authoritarian component of psychoanalysis. According to Federico Finchelstein's contribution to this volume, for instance, psychoanalysis was closely associated with anti-fascism. Finchelstein shows that, in Freud's mind, psychoanalysis constituted a transnational set of ideas that could oppose fascism (characterized as a transnational ideology). In Freud's view, by bringing repressed social atavistic vestiges of a barbarian past to the present, fascism could be understood as a social phenomena. Whereas in psychoanalytic thought myths have a metaphorical and analogical character, fascism took them literally as expressions of the soul. Psychoanalysis's rationality and emphasis on defenses and resistances of the psyche challenged fascist emphasis on the unbound will. Finchelstein thus makes a political reading of psychoanalysis which is, however, different from Schorske's classic interpretation. For the later, psychoanalysis was born as a result of a displacement of Freud's earlier political engagement into the self, turning political radicalism into an internal subversion of the self when participation in actual politics became difficult. In Finchelstein's view, psychoanalysis continued to have a deep political (anti-fascist) dimension. Thus, Gouda and Finkelstein point to different and seemingly incompatible appropriations of the Freudian system.

One distinctive feature of psychoanalysis is its multiple appropriations across different cultures. In post–World War II Europe, Gouda shows, psychoanalysis could be used to justify colonialism. If we consider also that psychoanalysis proliferated in Latin American countries precisely when those countries were ruled by murderous military dictatorships, then it becomes clear that, in opposition to Freud's ideas, there is some potential in psychoanalysis to become functional or at least passive when confronted with authoritarian societies and oppressive ideologies. Following Schorske, it could be claimed that in moments when the public arena is censored and repressed, psychoanalysis can provide a safe way to turn public interaction into a private negotiation with the self. Both victims and perpetrators of repression could benefit from the space of "privatization" of the public sphere opened by psychoanalysis. In a fascinating book based on a large number of interviews to former Argentine leftist guerrillas, María Matilde Ollier shows how, at least in that highly psychonalyzed society, psychoanalytic therapies, in many cases carried out in public parks and under fake names since patients and analysts alike shared their fear of being persecuted by the military authorities, provided important psychological support to leftist militants.[29] For the leftist militants it was safer for them to discuss their sexuality and dive into their inner self rather than engage in open political discussion and questioning of the existing social order. It is not by chance that neither in Argentina nor in Brazil, for instance, the military rulers did not repress psychoanalysis as such. In fact, in both countries, well-known analysts became conspicuously present in the state controlled media sometimes providing rationalization to the regimes' political ideology.[30] Of course, several psychoanalysts were imprisoned and a few of them were even "disappeared," but this repression was related to their political and human rights militancy and not to the fact that they were psychoanalysts. Exceptions notwithstanding, they were persecuted as citizens, not as psychoanalysts. Psychoanalysts were victims of the murderous Argentine dictatorship in a similar or even a smaller proportion than members of other professions.[31] It could be said, therefore that potentially, psychoanalysis has both liberating and oppressive dimensions that have to be explored cross-culturally.

While Gouda focuses on a "colonial appropriation" of psychoanalytical thought, Eli Zaretsky, Finchelstein and Elizabeth Danto look at the liberating potential of psychoanalysis. Finchelstein and Danto focus their attention on the context of origin of psychoanalysis, that is to say, in Europe. Taking as his point of departure the ironic dedication that Freud wrote on his book on war to Mussolini, Finchelstein explores the anti-fascist component of psychoanalysis. Danto, on the other hand, focuses her attention on the relations between modernism as a transnational "climate of ideas" and psychoanalysis, paying particular attention to the establishment of free psychoanalytic clinics that would liberate the working class from neurosis, in different European cities during the 1920s.[32] Like Finkelstein, Danto produces a political reading of psychoanalysis. In her view, psychoanalysis was associated not only with post-World War I cultural modernism but also with the modernist politics promoted by Austrian and German Social Democracy. Both Danto and Finchelstein focus on the usually forgotten political dimension of psychoanalysis as a theory and as a practice.

Zaretsky, on the other hand, concentrates on the uses and appropriation of psychoanalysis by black American intellectuals linked to the Harlem Renaissance. In particular, Zaretsky's chapter focuses on the works of Richard Wright who combined his readings of Freud with Popular Front ideology to make sense of the traumatic history and position of the blacks in the US. Psychoanalysis allowed Wright to get rid of the mood of lament and of the ideal of integration of blacks (to "move beyond the blues" in Zaretsky's words), and focus instead on resistance, mourning, working through of trauma and on helping blacks to give a voice to their self. According to Zaretsky psychoanalysis "exemplified the effort to give voice to traumatic suffering," effort that inspired Wright's works and thought. Whereas Danto and Finkelstein analyze the political aspect of psychoanalysis by focusing on the European context, Zaretsky emphasizes the transnational appropriability of psychoanalysis in a social and political context which was very different from its context of origin. There is a large scholarship on the history of psychoanalysis in the US, focusing on that country as a center for the diffusion of the discipline. Zaretsky's concentration on the reception of psychoanalysis among black intellectuals emphasizes what we could call "the periphery of the center."

Sergio Visacovsky, Alejandro Dagfal and Jane Russo study the transnational circulation of psychoanalytic ideas from a different perspective. By analyzing cases in which the diffusion of psychoanalytic ideas went from the periphery to the center, they implicitly question the "diffusionist" model of history of ideas that claim that ideas move from the centers of production to the periphery. In fact at least in two cases the history of psychoanalysis shows the opposite. Not only was Melanie Klein's British School of Object Relations known, discussed and practiced in Buenos Aires earlier than in Paris but, as Dagfal shows, the reception of those ideas in France took place through the works of Argentine analysts. Therefore, Klein's works became known in Paris through the publication of French translations of Argentine Kleinian psychoanalysts's works before her own books were available to the French public. The French reception of Klein's theories had, therefore, a distinctive Argentine flavor and French analysts became acquainted with the Argentine version of Klein before having access to Klein's works first hand. In this case the movement of ideas was south–north (or north–south–north) rather than the usual north–south. Dagfal also analyzes the psychoanalytic transnational network that began with the international mobility (voluntary or not) of analysts in the 1930s. This generated the material basis for the transnational circulation of ideas. Thus, French intellectuals who discovered psychoanalysis in Argentina became acquainted with Klein's works through other European analysts living in Buenos Aires who had traveled to England to receive training from Klein and her associates. The French–Argentine analysts' contacts in their home country allowed them to become a bridge between the Argentine, the English and the French psychoanalytic traditions.

Another clear example of the transnational circulation of ideas south–north is the case of the reception of Jacques Lacan's ideas in Spain, studied by Sergio Visacovsky. During the long Franco dictatorship in Spain (1939–1975) the formerly active psychoanalytic community of the Second Republic was silenced and dismantled.[33] It was only after Franco's death that Argentine intellectual and political exile Oscar Masotta (who has also been credited with the introduction of "Lacanism" in Argentina) re-introduced psychoanalysis in Spain, this time in the Lacan version. Again, in this case the circulation of psychoanalytic ideas was south–north rather than north–south. The paradox is that the Argentine Psychoanalytic Association created in 1942 was founded, and originally led, by a Spanish emigrée from the Spanish Civil War: Ángel Garma. Therefore, while IPA-affiliated orthodox psychoanalysis was introduced in Argentina by a Spaniard, Lacan's psychoanalysis was introduced in Spain by an Argentine. In both cases the "mediating circumstance" was the existence of a dictatorship in the country-source. Visacovsky also shows how the transnational circulation of psychoanalysis and people also implied a transnational construction of sources of legitimacy and genealogies. The name of Oscar Masotta, who died in 1979, was used by his followers and detractors to establish alternative genealogies for the Spanish and the Argentine Lacanian psychoanalytic movements alike.

Finally, Jane Russo studies the "Lacanian revolution" in two countries: Argentina and Brazil. Unlike Plotkin, who focuses on the early reception of psychoanalysis in both countries, Russo's chapter looks at more recent times. She carries out her analysis on two levels. On the one hand, she concentrates on the nature of Lacan's revolution of psychoanalysis. Thus, she argues that at the center of Lacan's version of psychoanalysis was an attempt to recapture the original "charismatic" nature of the discipline as opposed to the "bureaucratic" character promoted by the IPA. On the other hand, Russo shows how this process took place in Latin America and how the two countries she analyzes became transnational centers for the diffusion and practice of Lacan's psychoanalysis. In this case the circulation of psychoanalytic ideas and practice could be characterized as north–south–south, since the development of the Brazilian psychoanalytic movement has been very much influenced by Argentine psychoanalysis. Russo argues that the Brazilian "Lacanian invasion" was preceded and prepared by an earlier "Argentine psychoanalytic invasion" in Brazil.

At this point we could return to our point of departure and ask again why certain systems of thought (psychoanalysis in this case) become transnational while others fail? The essays that comprise this volume show that only a methodology that simultaneously focuses on the tensions and articulations between local levels of reception and international patterns of circulation on the one hand, and a comparative approach on the other can provide elements to answer this question. It is our hope that this book is a contribution in that direction.

Notes

1. Although Freud had started working on the construction of psychoanalysis before the publication of *Interpretation of Dreams*, this book has been widely acknowledged (even by Freud himself) as the foundational piece of psychoanalysis.
2. Since the 1920s Peruvian doctor Honorio Delgado had been corresponding with Freud and practicing his own version of psychoanalysis. See, Rey Castro, Alvaro, "Freud y Honorio Delgado: Crónica de un desencuentro." *Hueso Húmero*, 15/16 (January–March, 1983); and Rey Castro, "El psicoanálisis en Perú. Notas marginales." *Debates en Sociología*, 11 (1986). For the development of psychoanalysis in India, see Sudhir Kakar, *Culture and Psyche. Psychoanalysis and India* (New York: Psyche Press, 1997).
3. Carl Schorske, *Fin-de-Siècle Vienna: Politics and Culture* (New York: Vintage Books, 1981).
4. John Forrester, "A Whole Climate of Opinion: Rewriting the History of Psychoanalysis." In Mark Micale and Roy Porter (eds) *Discovering the History of Psychiatry* (New York: Oxford University Press, 1994), p. 174; and Forrester, *Dispatches from the Freud Wars: Psychoanalysis and Its Passions* (Cambridge: Harvard University Press, 1997), p. 2.
5. Sigmund Freud, *On the History of the Psychoanalytic Movement Vol. XIV of The Standard Edition of the Complete Psychological Works of Sigmund Freud*, ed. James Strachey *et al.* (London: The Hogart Press, 1991); and Freud, S., *An Autobiographical Study. Vol XX of The Standard Edition.*
6. Ernest Jones, *The Life and Work of Sigmund Freud* 3 Vols. (New York: Basic Books, 1953–1957).
7. Example of this are Jeffrey M. Masson, *The Assault on Truth: Freud's Suppression of the Seduction Theory* (New York: Viking Penguin, 1985); Even more bitter towards Freud is Peter Swales's "Freud, Mina Barnays and the Conquest of Rome: New Light on the Origins of Psychoanalysis," *The New American Review* (Spring–Summer, 1982), 1–23. An important attempt at demystifying the role of Freud, but still keeping him at the center of analysis is Frank Sulloway,

Freud, Biologist of the Mind (New York: Basic Books, 1979). See also Michael Roth (ed.), *Freud, Conflicts and Culture* (New York: Alfred Knopf, 1998). In recent years the "anti-Freudian" fashion arrived in France, see Catherine Meyer (ed.), *Le livre noir de la psychanalyse. Vivre, penser et aller mieux sans Freud* (Paris: Éditions des Arènes, 2005).

8. See, for instance, Schorske, *Fin-de-Siècle* (Chapter 4), and William McGrath, *Freud's Discovery of Psychoanalysis* (Ithaca: Cornell University Press, 1986).

9. Mariano Ben Plotkin's *Freud in the Pampas: The Emergence and Development of a Psychoanalytic Culture in Argentina* (Stanford: Stanford University Press, 2001), and Joy Damousi, *Freud in the Antipodes: A Cultural History of Psychoanalysis in Australia* (Sydney: University of New South Wales Press, 2005), for instance, explore the reception and evolution of psychoanalysis both as a therapy and as a cultural artifact in Argentina and Australia; Alexander Etkind's *Eros of the Impossible: The History of Psychoanalysis in Russia* (Boulder, Co: Westview Press, 1997) analyzes the early reception and diffusion of psychoanalysis in Russia; Nathan Hale's *Freud and the Americans* 2 Vols. (Oxford, UK; Hale: New York: Oxford University Press, 1970–1995); Eli Zaretsky's *Secrets of the Soul* (New York: Knopf, 2004) and Elisabeth Roudinesco's *La bataille the cent ans* 2 Vols. (Paris: Fayard, 1994) focus on the emergence and development of psychoanalysis in the US and France, among many others.

10. See, for instance, Zaretsky, *Secrets*.

11. See C.A. Bayly, Sven Beckert, Matthew Connelly, Isabel Hofmeyr, Wendy Kozol, and Patricia Seed, "AHR Conversation: On Transnational History." *American Historical Review*, 111 (5), (December 2006), 1441–1464.

12. Even the definition of who can be considered a psychoanalyst varies widely. While the IPA has established very strict regulations and requirements in this regard, Lacanians have established a completely different and much laxer system of training. See Roudinesco.

13. Jacques Alain Miller, Lacan's son-in-law has created an international Lacanian movement. However, among the Lacanians are those who dispute Miller's leadership.

14. Bayly *et al.*, "AHR Conversation."

15. On the migration of psychoanalysts from Europe in the 1930s see Riccardo Steiner, "It is a New Kind of Diaspora." *Explorations in the Sociopolitical and Cultural Context of Psychoanalysis* (London: Karnac, 2000).

16. Words originating in psychoanalysis have become part of everyday life in Buenos Aires, including the use of some neologisms such as the verb "psicopatear" or "histeriquear." See Plotkin, *Freud in the Pampas*.

17. Recognizing the weight of Latin American psychoanalysis, in 1993 the IPA elected Argentine Horacio Etchegoyen as its president.

18. See Meyer (ed.), *Le livre noir de la psychanalyse*, p. 7.

19. Ulf Hanners, "Flows, Boundaries and Hybrids: Keywords in Transnational Anthropology," WPTC-2K-02, Department of Social Anthropology. Stockholm University, www.transcomm.ox.ac.uk/working%20papers/hannerz.pdf.

20. For a discussion of these issues see Bayly *et al.* "AHR Conversation."

21. Zaretsky, *Secrets*; Berger, Peter, "Towards a Sociological Understanding of Psychoanalysis." *Social Research*, 32, (1965), 25–41.

22. Jean Paul Sartre said that while Marxism was the system of ideas of the twentieth century, only psychoanalysis could fill in its blind spot: its lack

of a theory of subjectivity. According to him, psychoanalysis "enables us to recover the complete human being in the adult, that is not only what he is now, but also the weight of his history." See Sartre, Jean-Paul, "Questions de méthode" in *Les Temps Modernes*, 13 (139), (September 1957), 380.

23. Zaretsky, *Secrets*, p. 6.
24. Peter Wagner, *A History and Theory of the Social Sciences. Not All That Is Solid Melts into Air* (London: Sage Publications, 2001).
25. Sherry Turkle, *Psychoanalytic Politics. Jacques Lacan and Freud's French Revolution* (Second Edition, London: Free Association Books, 1992), p. xiv.
26. Turkle, *Psychoanalytic*, p. xxiv.
27. Hannerz, "Flows."
28. "National habitus" is very broadly defined in Elias's work. However, we consider it a useful analytic concept. See Elias, Norbert, *The Germans. Power Struggles and the Development of Habitus in the Nineteenth and Twentieth Centuries* (New York: Columbia University Press, 1996).
29. María Matilde Ollier, *La creencia y la pasión. Privado, público y político en la izquierda revolucionaria* (Buenos Aires: Ariel, 1998).
30. In Brazil there is at least one documented case of an analytic candidate at the Rio de Janeiro society who had been a torturer. See Maria Auxiliadora de Almeida Cunha Arantes, *Pacto revelado. Psicanálise e clandestinidade política* (São Paulo: Escuta, 1994).
31. The diffusion of psychoanalysis during the military regimes in countries such as Argentina and Brazil is a phenomenon that still needs to be studied. For some preliminary hypotheses, see Plotkin, *Freud in Pampas*, Chapter 9.
32. See also Danto's *Freud's Free Clinics. Psychoanalysis and Social Justice, 1918–1938* (New York: Columbia University Press, 2005).
33. See Thomas Glick, "El impacto del psicoanálisis en la psiquiatría española de entreguerras." In Sánchez Ron, José Manuel (ed.), *Ciencia y sociedad en España: De la ilustración a la Guerra Civil* (Madrid: Arquero, 1988); Glick, "The Naked Science: Psychoanalysis in Spain, 1914–1948" *Comparative Studies in Society and History*, 24, (1982), 534–571.

Section 1

Psychoanalysis and Transnational Modernism

1

Three Roads from Vienna: Psychoanalysis, Modernism and Social Welfare

Elizabeth Ann Danto

If the transnational trends set by *début de siècle* psychoanalysts pick up again, and if analysts follow the tone of modernism set at their Congress in 1918, the twenty-first century should establish new records for advances in psychoanalytic thought and care. Nearly one hundred years ago in Budapest, Sigmund Freud himself led the way with a call for "the conscience of society [to] awake."[1] In Germany, Max Eitingon and Ernst Simmel rushed to implement Freud's Social Democratic covenant and to act on Freud's belief in achievable progress with, among other strategies, "clinics where treatment will be free."[2] The British psychoanalysts of the 1920s got caught up in the spirit too, though their modernism was not, on the whole, as openhanded as that of their colleagues across the Channel. Almost without exception the analysts' urban activism was located within and among nations where governments and citizens alike engineered deliberately new forms of social welfare planning. When we associate modernism with the history of psychoanalysis, we may perhaps envision a circle of intellectuals chatting at the Café Central, occasional incursions into art history, or even a sexually charged cinematic interpretation. Similarly, when the concept of transnationalism is coupled with psychoanalysis, we still respect the traditional distinct nation-state boundaries. This essay, then, aims to shift this ambiguous narrative from psychoanalysis as a lone clinical construction, to psychoanalysis as a modernist social welfare ideology, born of *début de siècle* Vienna and bred as a flow of ideas and practices across geographies.

"Indefatigable eagerness"

From 1920 to 1933, arguably the most generative period in the history of psychoanalysis, members of the International Psychoanalytic

Association (IPA) formally refuted Europe's monarchist traditions, not only with their belief in the dynamics of the individual unconscious, but also by pooling their creativity toward the greater good: they joined municipal governments, mounted lecture programs in the public schools, advocated for reforms in health and mental health, planned clinics (some brought about and some not) for indigent citizens of Vienna, Berlin, London, Budapest, Zaghreb, Moscow, Frankfurt, Trieste and Paris. Each clinic treated roughly 300 patients yearly at no cost, according to figures collected by the *International Journal of Psychoanalysis*, offered free training analyses to institute candidates, and were well-liked by popular news magazines. "Lonesome: You are 29 years old, intelligent, educated, with a good job and you long for a companion who would share with you sorrow and joy," wrote an advice columnist from *Bettauer's Wossenschrift*, in 1924. "This is no doubt a case which necessitates psychoanalytic treatment. Consult with the Psychoanalytic Ambulatorium, Vienna, Ninth District, 18 *Pelikangasse*. Intake from 6 to 7pm."[3] The Ambulatorium was the free psychoanalytic clinic in Vienna, supported directly by Freud and his colleagues. Had the analysts kept up this pace for the remaining decades of the twentieth century, their collective effort would have caused an immense ideological swell – not enough to prevent the Nazi's decimation of their physical spaces – but certainly powerful enough to be re-narrated across nations, as an achievement of European modernism, one akin to Schoenberg's atonal music and Adolf Loos's ruthlessly streamlined architecture of the era.

By chance or intention, the psychoanalysts of the 1920s tapped into a powerful cultural longing: the yearning of thousands of sensitive people determined to establish a new form of citizenry and find political solutions to the unfamiliar, post-war (and post-monarchy) social crises. And unlike the earlier cultural expansions signaled by Monet's Impressionism and Nietzsche's "Will to Power," both of which dismissed positivism's claims to certainty as romantic fantasies, members of the IPA had a physical location, compelling leadership, and a commanding passion for the revolutionary quality of a body of theory. "Our glorious esprit de corps and... the indefatigable eagerness with which we devoted ourselves to the 'cause,' gave an exhilarating feeling... We experienced vivendo what Freud said about the power of such participation: 'Men are strong, so long as they represent a powerful idea'," wrote the psychoanalyst Richard Sterba.[4] Breaking with the pre-war idea of a local geographic center in Vienna, the psychoanalysts adopted the plan of other transnational movements like Surrealism, Cubism, Bauhaus and even Leninism, all movements that spread purposely beyond their original base. The analysts' group was, after all, "International" in name and in scope virtually from the beginning. The clinics were the practical implementation of this transnationalism. They were free clinics literally and metaphorically: they freed people of their destructive neuroses and, like the municipal schools and universities of Europe, they were free of charge. Psychoanalysis would share in the transformation of civil society and these new outpatient treatment centers would restore people to their inherently trustworthy and self-regulating selves. As psychoanalysis spread, Europe's intellectuals built over earlier theories or used them as cautionary signposts. Anna Freud, who grew up with them, said that psychoanalysis was seen "after the First World War and in the early '20s as the embodiment of the spirit of change, the contempt for the convention, freedom of thought about sex and, in the minds of many, the eagerly looked for prospect of release from sexual restrictions."[5]

If modernism is defined as the twentieth century's expansive social and cultural production that sooner or later involved every art form and government, then early psychoanalysis was as modernist as Schoenberg's *Second String Quartet* and the Social Democrats' redistributive taxation policy.[6] At the same time, the psychoanalysts articulated an overarching transnationalist worldview; the political implications in doing so were that it could alienate them from established academia and medicine just when the movement was becoming widely known. Yet the movement could already be mapped out transnationally, on a continuum in which Vienna's Wilhelm Reich represents the expansionists, London's Ernest Jones embodies the idealism of neoconservative local autonomy, and Berlin's Max Eitingon lands ideologically between the two. Perhaps no thinker was more influential in this arena than Sigmund Freud whose writings like "Civilization and Its Discontents" formed a selective encyclopedia of human ingenuity. Freud actively supported Europe's progressive post-war reforms and locally, enjoyed contemporary recognition from Vienna's liberal newspapers. Who else, asked the *Presse*'s editor, could "be found who will possess the same high degree of awareness of their social obligations as this internationally known Viennese scholar?"[7] By then Vienna was a newly constituted autonomous state, and the *Presse*'s readers had voted in the Social Democratic Party to build a new kind of post-war government, one based on a redistributive vision of the future, not on current constraints. Freud (like Jones and Eitingon) maintained a calibrated approach to the Armistice, but he did write to his great friend Sandor Ferenczi that the Hapsburgs "left behind nothing but a pile of crap."[8] Although Ferenczi would not be considered as one of the psychoanalytic movement's experts in social activism, few others spoke as comfortably as he did about the *"real* conditions of the various levels of society." In fact, contrary to today's conventional wisdom about the class-based history of the psychoanalytic movement, virtually all of the analysts of the 1920s, from Reich to Erich Fromm and Erik Erikson, did so on a surprisingly broad scale. "We were rebels, in our own ways," said Grete Bibring, a Viennese psychoanalyst who studied alongside Reich in medical school and taught at Harvard after Hitler's takeover. "We stood with the poor, and wanted to fight for their interests. For us psychoanalysis promised personal 'liberation' not for its own sake, but so that we could work to 'liberate' others. Political social and activism, they were a big part of our lives."[9]

Though nearly a century has passed, still we rarely think of psychoanalysis as attending to blue-collar workers, government functionaries, public school teachers, farmers, factory workers, poor children and unemployed parents. Perhaps the analysts' assault on nineteenth century class structure (that everyone has an unconscious, that women have sexuality) prompted this social resistance. Critics, especially Americans, suggest that individual psychological investigation precludes environmental advocacy and that psychoanalytic studies move the individual person away from culture.[10] Empiricist intellectuals enjoy invalidating psychoanalysis as non-scientific and purely ideological. Psychoanalysts themselves have alleged that clinical objectivity demands distance from politics, social policy and social thought. Even within the movement Wilhelm Reich, always alert to a group's internal contradictions, noted that "the conflict within psychoanalysis in regard to its social function was immense long before anyone involved noticed it."[11] At his most biting though, Reich still saw that his colleagues had embarked on a far-reaching corrective strategy. Reich, arguably the most original of psychoanalysis' many *enfants terribles*, took the movement to its further modernist potential.

"For the wider social strata"

Few cultural forces have so clearly shaped the future of Western urban populations as modernism, and psychoanalysis matured along with it. The analysts' sympathy for the victims of the recent war (but not the perpetrators) extended beyond nationalist boundaries. Thus, not only were uncounted numbers of individuals relieved of personal anguish; so were the dozen or more city governments where psychoanalysts installed free clinics and helped rationalize social welfare on a larger scale. On the personal or individual level, psychoanalytic practice that was 95 percent unconscious (or virtually unknowable) and 5 percent conscious (visible and knowable), met modernism's standard of anti-positivism. Dispensing with mid-nineteenth century conventions of *realpolitik* and traditional social rhythm, *début de siècle* psychoanalysts adopted a complex new form of human exchange, and rearranged traditional power assumptions in patient–physician relationships: in sitting behind the couch and out of the patient's sight range, psychoanalysts listened to the narratives in the room. In this way the analysts reversed the traditional icons of cure where the omniscient prescriptive doctor instructs the supplicant patient; harnessing the intensity of this major shift was a signature of progress. But modernism was not only a reaction to the historical past, and it championed a philosophy in which social forces and human rationality were more important than facts or material goods. Thus, by its standard of aesthetic rationality (sexuality stripped of euphemism, relationships without sentimentality), psychoanalysis was utterly purposeful: with all the dreams and free associations a patient funnels to an analyst, and with all the interpretations the therapist reflects back, the objective of the process is to free patients of their life impediments. Where the patient's heartfelt drama centers perhaps on family (where relationships may be dominated by an earlier oppressive hierarchy), analysis draws on the patient's capacity for self-regulation obscured until now.

But individual treatment alone never satisfied the demands of modernism's futurist political scope nor the psychoanalysts' re-examination of concepts like citizenship, nationalism and community – the very same concepts underlying transnationalism. Europe's housing shortages and post-war urban dislocation challenged the analysts to fight for fair government, and win. Changing the world was not impossible, but it required belief in a collaborative future and unsentimental rejection of the past. In this way Freud, who held the idea of achievable progress in high esteem, and who believed that any unknown aspect of the world required rational inquiry, urged the analysts onward. Sixty-two years old and frankly impatient with the old idea of the absolutist state, he wrote to Ferenczi that "the stifling tension, with which everyone is awaiting the imminent disintegration of the State of Austria, is perhaps unfavorable." But, he continued, "I can't suppress my satisfaction over this outcome."[12] Even before war's end, Freud's September 1918 address to the Fifth International Psycho-Analytical Congress charted the movement's future. In by far one of his most eloquent speeches,

Freud appealed for post-war social renewal on a transnational scale, a three-way demand for global civic society, government responsibility and social equality – an appeal that reached well beyond the relationships between and among individuals within a single nation-state. Had he been heard by a larger audience, he would have set off a national debate, not just over the central role of government but over the range of questions on the nature of social progress – the need to reduce inequality through universal access to treatment, the influence of environment on individual behavior and the necessity of modernization.

Freud's argument concerned nothing less than the complex relationship between human beings and the overarching social and economic forces. Implicitly he was throwing in his lot with the local Social Democratic government (the party of modernism in Vienna) but also projecting a global borderless scope for psychoanalysis. In some extraordinary work promoting the progressive agenda – one rarely associated with psychoanalysis today – Freud mapped out a form of social welfare planning that would last, in practice from 1918 through 1938, and in concept far beyond. The result placed psychoanalysis within modernism's specific intent not to duplicate the institutions of bourgeois society. Likewise, Freud unequivocally believed that confining the movement to national boundaries would preempt the flow of psychoanalytic ideas and services. Thus the free clinics were community-based at the local level, all the while building transnational support. In Vienna, the analysts' local contributions to social welfare corresponded, patient by patient, to the political returns received from Julius Tandler, the great Austrian anatomist who, in 1921, put public welfare into action and laid out professional social work as we know it today. In the city dubbed "Red Vienna," Tandler revamped health, housing and family policies from patronizing selective charity to universal social welfare; as Freud stressed in 1918, "the poor man should have [the] right to assistance."[13] How many free clinics the analysts developed (we know of roughly a dozen, but more may have been built) depended on the depth of their convictions that, as Freud also said, "It may be a long time before the State comes to see these duties as urgent...Probably these institutions will be started by private charity. Some time or other, however, it must come to this."[14] The mission was located deep within social consciousness without which psychoanalysis risked marginalization. Until 1918, a fervent anti-traditionalism (and its largely Jewish membership) had driven the psychoanalytic movement outside time-honored medical and academic communities. Now its survival depended on a new governmental configuration, one where the mental health of its citizens was a state responsibility. In a series of ideological positions intended to de-stigmatize neurosis, Freud sought to redefine a personal trouble into a larger social issue, and spreading outward, to make the transnational civic community accountable for the care of individual mental illness.

The argument for this kind of transnationalism intensified with every new step of the psychoanalytic movement. Like his friends and contemporaries, Freud believed that measurable social progress would be achieved through a planned partnership of the state and its citizens. The power to set a country's laws should be redistributed democratically to its citizenry. Citizens had social rights, among them health, housing and welfare for children and worker's families. An interventionist government would forestall the increasingly obvious despair of overworked women, unemployed men and parentless children. The political and social gains derived from the psychoanalysts' new alliances would, at the very least, confer legitimacy on a form of mental health treatment often practiced by non-physicians, or by physicians reluctant to join the establishment. Otherwise, psychoanalysis would stay stuck in positivism and the analysts' lack of social awareness would render them virtually powerless, doing "nothing for the wider social strata, those who suffer extremely from neuroses."[15] In 1918 Freud might have simply restated the 1913 principles that systematized his pre-war approach to patient fees. But the advent of modernism in politics (as in music, architecture and art) demanded greater public involvement and accountability all around. In turn Freud challenged his colleagues to attend collectively to their social obligations. The Budapest speech on "the conscience of society" reflected Freud's personal awakening to the reality of a new social contract, a new cultural and political paradigm that drew in almost every reformer from Adolf Loos in architecture to Clemens Pirquet in medicine and Paul Lazarsfeld in social science.

"A revolution in the soul of man"

Much of this transformation occurred not in Austria but in Germany where Max Eitingon and Ernst Simmel, two of the movement's great unsung champions, shaped the moving force behind the "Berlin Poliklinik." Actually Eitingon had already financed a rudimentary independent psychoanalytic service in 1910; after the 1918 Budapest Congress, this turned out to be the blueprint for the Berlin Poliklinik, the first free psychoanalytic clinic. The Poliklinik, as it came to be known, opened in 1920 and represented a highly successful, years-long effort to change the public image of the psychoanalyst as physician/ entrepreneur into one of social reformer at the service of indigent people. For Eitingon and Simmel, the Poliklinik's presence in Berlin could convert the tenuous, even outright antagonistic, relationship between psychoanalysis and clinical psychiatry (as between individual mental health and the government) into collaboration. There was a precedent: Germany's urban psychiatrists had started polyclinics in the 1890s to offset their own profession's poor public image. Thirty years later, the Poliklinik launched a similar mission but added fresh approaches initiated in other cosmopolitan centers. Melanie Klein's child analysis and Ferenczi's short-term treatment and crisis intervention for example, came up from Budapest.

As for Max Eitingon, he looked to be, by the end of 1920, a lot more than another Freud partisan. He joined psychoanalysis in 1905 after studying psychiatry with Carl Jung at the celebrated Burghölzli clinic in Zurich. After 1918, Eitingon hoped to advance the debate on progressivism in psychoanalysis while broadening the movement, but he had found himself captive to the inner "Committee" and especially the internecine mischief of Ernest Jones. He disliked this sort of Oedipal argument, and one senses from reading his correspondence that, if the Poliklinik also proved to be his major accomplishment, he would survive the disappointment and it would not be his last. Then again, he made new friends. With the avowed socialist Ernst Simmel at his side, Eitingon helped the analysts came to terms with Griesinger and the stars of traditional psychiatry at Berlin's Charité. Though he had entered transnationally as a psychoanalytic iconoclast (the commentariat of Rado was that Eitingon was, at best, a good administrator), it was Eitingon who, inspired by Freud's 1918 call to modernism, transformed the psychoanalysts' treatment community into a democratic auxiliary by insisting that colleagues support a broad agenda of clinical experimentation, treat people of every occupational status and diagnosis and contribute a small portion of their designated fees to the clinic's upkeep. He was one of Freud's most ardent, though personally dispassionate supporters, and used his extraordinary personal wealth for one purpose – to guarantee access to respectful mental health treatment to indigent people who had no contact with psychiatry beyond state institutions.

Eitingon's brilliant stroke of February 1920 was staging a day-long cultural event which, while overtly the formal inauguration of the clinic, may have forced a shift in perception of the nature of psychoanalysis as a modernist paradigm, itself imbued with transnationalism. The program was made up of four sections held together by a common thread – the relationship of innovation to tradition – and the parts were idiosyncratic enough to comment on one another like the multiple sides of a Klimt figure. A performance of Chopin and Schubert chamber music registered as intimate and cultivated, while on another plane the day's series, which ranged from a Schoenberg piece and art songs by Hugo Wolf, to a reading of "Presentiment" and "Madness" from Rilke's *Book of Hours*, was high-concept modernism. The program was as psychologically blunt as modernism could be yet the tone, at times, verged on a broader classicism with its overall symbolist themes of human emotion, reality and nature. Traditional pieces from the mainstream of German culture were paired with contemporary work depicting modernity and subjectivity. Schoenberg, for example, identified musically with the Expressionists and politically with the Social Democrats. In poetry, the psychoanalysts offset Rilke's romantic voice with the biting surrealism of Christian Morgenstern's satire. Rilke was still living in Europe then, enormously popular though still edgy, and like Freud, an intimate of the Russian intellectual Lou Andreas Salome. In its own way the Poliklinik's inaugural program was a dialectical version of the analysts' views on the scope of modernism and their place within it.

Whether it was because Freud experienced a real political conversion in the course of the war, or because others like Otto Fenichel and Ernst Simmel had always responded to alternative social arrangements, the rejection of functionalist standards in favor of a new activist paradigm seemed remarkably natural. The reach of psychoanalysis would be global; in a sense, it was the post-war economic uncertainty that would make possible the construction of a worldwide network. Berlin's progressive intellectuals applauded the growing reach of psychoanalysis into new Weimarian forms of media – theater, cinema, photo-journalism and poetry. "The philosophy of the unconscious, initiated by Freud [grew] daily in Berlin," recalled the theologian Paul Tillich.[16] When Freud urged "individuals or societies... elsewhere to follow Eitingon's example, and bring similar [clinics] into existence,"[17] he hoped to persuade his international colleagues, including Abraham Brill in New York, to attract similar audiences. Freud reasoned that if Brill visited the Poliklinik, with Melanie Klein's child analysis and Franz Alexander's frequent lectures on theory, he would help American psychoanalysis escape from the nationalist dollar-bound sway of the "Progressive" Movement.[18] One way or another, in the mid-1920s, with Berlin taking hold in the interstices of transnational culture, the Poliklinik became the psychoanalytic version of the Bauhaus teapot: humanist, utilitarian, non-traditional, urban and as innovative in form as in content.

The psychoanalytic clinics appear at first glance to be born in the context of individual nation-states. One can easily change lenses, however, and view them as a borderless network, a coordinated group of institutions based on common clinical references and common political interests among members. Resources, information and decision-making flowed across national borders to uphold the clinics' larger social welfare purpose. Some of Europe's most aggressively modernist artists argued that cities would pass a decisive point into the twentieth century once they were able to continuously innovate in architecture, music and literature, because this would show their mastery of social democracy. In Vienna for example, Adolf Loos (the architect for whom tradition undermined progress in all areas of culture) commented that, while he was living in the twentieth century, his neighbor was living in 1880 and some of the peasants in the Austrian provinces were still living in the twelfth century.[19] Similarly, though the *Allgemeines Kranken-haus* (the bastion of establishment medicine) still supported Theodor Meynert's unfashionable legacy of strict empiricism and Julius Wagner von-Jauregg still dismissed psychoanalysis with contempt, the transnational force of modernity compelled them to face de-territorialization in psychiatry. In 1922, the same year Vienna celebrated its voluntary independence from Lower Austria (the *Trennungsgesetz*, or Statute of Separation), Wagner-Jauregg's governing "Society of Physicians," a formidable group of conservative neurologists and psychiatrists, agreed to let non-physicians treat mental illness. They gave the analysts the use of hospital facilities for an out-patient center and officially chartered the clinic along the prevailing lines of free medical care, free education and low-cost housing already in place.

Whether in Germany, Austria, England or Hungary, the clinics had one clear advantage over old line psychiatry: it was integrated into the post-war emerging social service networks that, with their dynamic mix of physical and mental health programs, hastened the growth of the economy. In Vienna paradoxically, the multiple social services eventually built into the city's modernist community housing complexes, the celebrated *Gemeindebauten*, reinforced an interesting contradiction between the state's genuine assistance to families (in order to grow healthy children) and, simultaneously, the maternalist limitations this placed on women's lives (the demand to procreate). By their own admission, the psychoanalysts like Martin Pappenheim (a left-leaning neurologist and a frequent guest of the Freuds) also maintained that social change should reach "into the structure of family relationships, the social position of women and children, [and] sexual reform."[20] The current social movement was to be a revolution in "the soul of man."[21] Urban culture, as Otto Bauer said when editor of the socialist journal *Arbeiter-Zeitung*, should reach from the privacy of individual and family life to public policy and the workplace, encompass the worker's total life.

This way of thinking about a mental health clinic, in Berlin and in Vienna and later in Budapest, clearly sought to rebuff earlier methods of psychiatric practice, and not just by adjusting past knowledge in light of current psychoanalytic technique. Traditionally, mental health care had been provided locally, in oppressive state facilities such as Vienna's sprawling *Am Steinhof* and the *Landesirrenanstalt* (State Lunatic Asylum) where Julius Wagner-Jauregg systematically tried to cure mental diseases, mostly the psychoses, by inducing fevers and shock. At Berlin's Charite, Wilhelm Griesinger decided which psychiatric patients would be restrained and which would not. Freud, as head of the Vienna Psychoanalytic Society, made it his business to upset that kind of narrow, and nationalist, authority over patients. He and his colleagues testified against Wagner-Jauregg's use of electroshock in court and fought back the medical establishment's two-year effort to close the Ambulatorium. On a theoretical plane, they argued that, unlike an external and objective reality in which traditional physicians believed, people's experience of reality is actually subjective and is based on a continuously changing (and unconscious) interplay between mind, body and environment. In this way at least, the mind of a psychotic person is no different from the body of a person with physical ailments, and deserves equally respectful and individualized treatment. Modernism in psychiatry meant out-patient clinics, the private office restructured in concept and in space as a community. In other words, the challenge at both the Poliklinik and the Ambulatorium was not so much making sure that poorer patients got private care, but that individual care in itself could be justified ideologically. As Ernst Simmel would argue from Berlin in *The Socialist Physician* of 1925, the paradigm of individual patient/individual physician is the stamp of bourgeois medicine regardless of its geography (private office, locality, nation). In contrast the paradigm for socialist medicine is the group: the cross-pollinating medical team, the waiting room as the patient community. This cultural heterogeneity is impossible to achieve without a structural reorganization of the global health care system. At the very least, a clinic would be the start of building on the new paradigm.

"Frighteningly thrilling"

"It is good that the old should die, but the new is not yet here," Freud wrote Eitingon a few weeks just before the end of World War I; his first glimpse of freedom from war was "frighteningly thrilling," and he and his colleagues had a very real sense that remarkable changes at all levels of society were about to transform the world they had known.[22] Before the war, the psychoanalysts had paid relatively little attention to political movements. But that changed in 1918, in large measure because they saw how Victorian positivism, with its questionable connotations of truth and practicality, had not prevented war, arguably the ultimate upsurge of aggressive drive. World War I was a cataclysmic failure of the earlier political status quo, and the post-war generation stressed that society had to make a deliberate effort to refrain from killing millions of people over scraps of land. The main reason for the public's senseless acquiescence to war, they figured, was that rational argument (as opposed to impulsive violence) is a long-term, costly matter that requires periodic state assistance, one in which the governing Hapsburgs traditionally had little interest. But those who had systematically challenged previous ideas – the nascent modernist architects (Alfred Loos, Ernst Freud), composers (Berg, Von Webern) and writers (Thomas Mann, Bertold Brecht) of the era – drew other conclusions, and their contemporaries involved in psychoanalysis could explain the new ideas as well. For Freud, our way of understanding the mind (both individual and social) would have to shift from the linear organic-biological perspective to a dialectical view of mental processes, an unconscious system of fundamental impulses and counterbalancing restrictions, a system that appears irrational but can, in fact, be explained through a rational lens. Not that the war should be excused any more than other superego failure, but the war-mongering behavior did deserve an explanation. Explaining the unexplainable was key to the modernist worldview.

As the painters, architects, composers and writers pursued the character of modernism with the idea of replacing establishment aesthetics with a new system based on popular culture (that is, derived not from high culture or self-conscious theory, but instead from the actual world including cross-border mass industrialization) so too did the psychoanalysts. The psychoanalytic clinics quickly became a platform for new treatment methods specific to contemporary urban demands, to search for new results, to try out analysis with children and older people, and to experiment with variations in the analytic hour, the duration of treatment, even to share responsibility with the patient. Is the therapy hour 60 minutes or 45, or could it vary? Can individual treatment be shortened by weeks, months or even speeded-up? Are such decisions best made by the patient or the clinician? These moral and practical controversies were argued often though inconclusively by the clinicians. At the clinics, and especially in Berlin, the analysts rejected expedient solutions to human problems whether in psychotherapy, culture or later, the looming crises of fascism and economic depression. Some took an expressly leftist political stance: Wilhelm Reich and Edith Jacobsohn joined the Communist Party, Erich Fromm and Otto Fenichel were declared Marxists, Barbara Lantos and Frances Deri were Socialists. Even without joining a specific political party, Grete Bibring recalled, "many of us were politically very active. We were socialists, or 'liberals' who wanted to change the society."[23] In the community or among the psychoanalysts, some of the quandaries (whether economic or social, internal or external) brought on by this stance were solved with barter. The training analysis was a case in point. Because the clinic analysts were required to have had a personal analysis (a stipulation agreed on at the Budapest conference in 1918) and most had been no-fee candidates themselves, they were expected to reciprocate. Even in the last days of the Vienna Institute before the 1938 *Anschluss*, "every doctor had non-paying patients... and every training analyst treated two candidates free,"[24] recalled the psychoanalyst Else Pappenheim who later practiced in New York. This form of exchange might be "surprising to Americans today," she said, but the same anecdote has been repeated at different times, in different contexts, by Eduard Hitschmann, Grete Bibring, Richard Sterba and Helene Deutsch. As Freud said bemusedly to his friend Franz Alexander in Berlin, "Almost all the training analyses in Vienna are carried out gratis."[25] These ambitious young left-wing analysts who staffed the Ambulatorium, for whom traditional forms of exchange hindered social progress, were not particularly radical for Red Vienna. They were, however, members of a small society whose product led the way toward the larger society's more generous view of human possibility.

Throughout the early years of the free clinics, psychoanalysts in various countries followed a well-organized sequence, a "logical order" Ernest Jones would say, of first constituting a local society among themselves, then issuing a clinical journal, and third organizing a training institute either before or just after opening the clinic. In London, Ernest Jones sought to "consolidate the new profession of psycho-analysis" by fusing the Society, the institute and the clinic into a single unit.[26] The most original element of the project – Jones called it "revolutionary" – placed lay analysts at the clinic in the same capacity as the regular staff. Interestingly, in London, Berlin and Vienna, though all at different times, new clinical knowledge was discussed at the Society but implemented only at the clinic. This shift in power relationships, paralleling similar shifts in the larger national ideologies, allowed the psychoanalytic clinics to take in hundreds of patients. Eduard Hitschmann, a Viennese psychoanalyst who belonged to Freud's inner circle and ran the Ambulatorium from 1922 until 1938, reported that schools and clubs, teachers, school-doctors and personal pediatricians, in fact all municipal social welfare agencies freely referred children "from all strata of the necessitous classes" to the clinic. And in Berlin at the end of the 1920s, at least according to Sándor Rado, the Poliklinik clinicians conducted "110 free treatment analyses everyday," not including the training analyses.[27]

Paradoxically, the same incrementalism of social democracy that forged a common ground among the senior analysts, disappointed a number of younger practitioners. A psychology of dialectical materialism, as Otto Fenichel would call it in 1934, offered a more radical (and necessary) alternative to the IPA's varying response to bourgeois influence. For Erich Fromm, Ernst Simmel, Annie and Wilhelm Reich, Edith Jacobson, Francis Deri, Bertha Bornstein, Kate Friedländer, Siegfried Bernfeld, Alexander Mette and Barbara Lantos, the age of modernism exacted both urgency and idealism. From November 1924 until at least 1933, this "special group [that] tackled the relations between sociology and psychoanalysis," as Edith Jacobson said, gathered 168 times in groups of 5–25, as a semi-formal study group that met every few weeks in private homes. Their enthusiasm for wide-scale social change was so intense that it left little room for alternatives to psychoanalysis beyond the conviction that, as Fenichel would say later from exile, "we recognize in Freud's Psychoanalysis the germ of the dialectical-materialist psychology of the future, and therefore we desperately need to protect and extend this knowledge."[28]

The real achievement of the modernist stance in psychoanalysis was not that, as Ernest Jones wrote to Freud a decade earlier, "in your private political opinions you might be a Bolshevist, but you would not help the spread of Ψ to announce it." [29] On the contrary, it was a transnational intellectual adventure, a persistent urge to confront the impingements of traditionalism, capitalism, and soon fascism, on human freedom. Psychoanalysis could – and should – have genuine meaning for the proletariat, but only if it is put to practical use in the class struggle. Otto Fenichel persevered with this argument, one made even more poignant and provocative after his multiple escapes from fascism and his final exile in Los Angeles. Fenichel was a thoroughgoing modernist, to some a portrait in contradictions and to others a champion of the unconventional: he challenged classical psychoanalytic technique in Berlin, then wrote the three-volume textbook of psychoanalytic orthodoxy, *The Psychoanalytic Theory of Neurosis*; partnered with Wilhelm Reich to promote dialectical materialism in psychoanalysis, then excoriated Reich for his radical position; campaigned against the American Psychoanalytic Association for its "MD only" policy of the 1940s, then died trying to gain American medical credentials.[30] Ultimately though, Fenichel was unshakably faithful to a single ideal – human liberation – and was, in turn, crucial to its development.

For the psychoanalyst Siegfried Bernfeld, the very picture of transnationalism in psychoanalysis, who traveled back and forth between the cities of Europe throughout the 1920s, bourgeois hegemony was his generation's monarchy. Bernfeld was an incredibly energetic and influential psychoanalyst, educator and activist, who intersected with the worlds of Fenichel, Freud, Wilhelm Reich, Walter Benjamin, Martin Buber, Ernst Simmel and so many others. He was amazingly productive and, throughout his life, dedicated to virtually any cause that would liberate people from oppression. For youth, his leadership of the Anfang Movement sought a voice independent of parental power; for students, he wanted a less authoritarian education; for people with mental illness, he built up free psychoanalytic clinics. He was less intellectually abstruse than Benjamin, more overtly political than Freud (though a great favorite of his, as well as of Anna), less spiritual than Martin Buber while an adamant Zionist, and as zealous a liberationist as Reich. Bernfeld's contributions to revolutionary cultural theory were critical and his modernist mission was unswerving: as long as medicine (or, for that matter, all hegemonic institutions) remained largely in the private hands of the bourgeoisie, stonewalling both socialism and psychoanalysis' efforts to build up the nation's health, crisis was inevitable. Whether it took the form of Fascism (as Benjamin and Simmel argued) or self-destructive rigidity (according to Reich), the result would be an ignoble decimation of human potential. Somehow the bourgeoisie would have to be compelled to restructure health and mental health care, redirecting it toward the working classes. Bernfeld's friends in Berlin and Vienna agreed. In 1931, Walter Benjamin wrote to the playwright Bertold Brecht that the purpose of his new journal *Krisis und Kritik* (to which Reich was an invited contributor) was to articulate the doctrine of "dialectical materialism by applying it to questions that the bourgeois intelligentsia is compelled to recognize as its very own."[31] Bernfeld argued that the two streams of psychoanalysis (the dialectic of the theory and the practice) are equally powerful and together have a political impact that neither element alone can achieve. Even family conflict and inter-group tension – and, by extension, transnational tension – previously viewed within a traditional individualist framework, could be brought to bear on the class struggle.

Among the politically engaged analysts who worked with Bernfeld was Ernst Simmel, a former army psychiatrist who co-directed the Berlin Poliklinik with Eitingon. Along with Reich and Fenichel, Ernst Simmel was one of the movement's most ardent social activists and, by starting Schloss Tegel, the first in-patient psychoanalytic facility, a defining voice in *début de siècle* modernism. He was an early promoter of psychoanalysis for people with categorical psychiatric disorders, edited *Der Sozialistische Aerzte* (The Socialist Physician) whose contributors included Albert Einstein and Käthe Kollwitz, and believed, deeply, that the "fundamentally egalitarian nature of psychoanalysis" (Simmel, 1930) would alter class-based inequalities in clinical practice. Like his contemporaries in music, art and cinema, Simmel held that necessary social advances required people to overcome their resistance to change on a very broad scale while, perhaps counter-intuitively, also standing firm against totalitarianism. The Russian Revolution had catalyzed his avowed socialism, a kind of synthesis with functionalism that one might call modernist Marxism also seen in the works by Brecht, W. H. Auden, Andre Breton, Louis Aragon as well as Gramsci and Walter Benjamin (not that all intellectuals of the 1920s were of the left; Wyndham Lewis, T. S. Eliot and Ezra Pound were serious conservatives). In the early 1920s, Ernst Simmel was simultaneously awarded the chairmanships of the German Psychoanalytic Association (*Deutsche Psychoanalytische Gesellschaft*, or DPG) and Berlin's Socialist Physicians Union. The Union's study groups explored legalizing the eight-hour work day (along with its health implications and cultural meaning), occupational health and safety, maternity leave for pregnant and nursing mothers, child labor laws, and socialized medicine. They fought for birth control and against the criminalization of abortion. Along with Bernfeld and Fenichel (as we have seen, two of the most politically dynamic psychoanalysts), Simmel argued that theory and practice together – praxis – had a political impact that neither element alone could achieve. This praxis would push the state to collaborate with medicine and psychiatry, and to redirect its considerable resources toward a more global health care governance, one less specifically limited to boundaried specializations. This synthesis was the ideological underpinning of Schloss Tegel, Simmel's ambitious second psychoanalytic enterprise, a sanitarium just outside of Berlin where indigent people with severe psychiatric disorders would be treated psychoanalytically. Simmel created a total therapeutic environment with a core dialectical materialist understanding of the patients' often-turbulent behavior, one where psychoanalysis could be used "for the relief of those patients whose extremity is greatest and who hitherto have been condemned to death in life."[32] Simmel re-introduced short-term psychoanalysis from his wartime work with shell-shocked soldiers, and combined "analytic-cathartic hypnosis with analytical discussion and interpretation of dreams," that, he said, "resulted in liberation from symptoms in two to three sessions."[33] Eventually the strategy of open psychiatric facilities, where patients could walk out of the building independently and roam around the castle park accompanied – or not – by their analyst or nurse, made for strained relationships within and outside the facility. Nevertheless, Simmel's appreciation for the role of community in psychiatric recovery, the cultivation of the sense of belonging so necessary in the evolution of a transnational identity, deepened his goal of partnering clinical intensity with a specific architecture.

"The influence of the environment"

Particularly at the beginning, Simmel found, the struggle to broker a new social arrangement between patients and their environments required an architect with a sure foot in modernism. Ernst Freud came at the recommendation of Max Eitingon, for whom the young architect was systematically redesigning a large apartment on Potsdammerstrasse, a suite of six high-ceilinged rooms rented for what would become the treatment and conference rooms, lighting and furniture of the Berlin Poliklinik.[34] Ernst had recently moved from Munich to Berlin, Germany's two poles of architectural ferment, and was developing his own functional architectural style mid-way between Adolf Loos's unadorned, controlled spaces and Richard Neutra's environmental domesticity. By the 1920s, most European countries had legislated forms of urban planning predicated on technology, public housing and public transportation, and architects grappled with new problems of high-density apartment dwelling. Inevitably, political ideology played a role. As Gaetano Ciocca, the Italian engineer who made his reputation on the fringes of *début de siècle* political paradoxes, wrote about contemporary architectural challenges: "Under pure capitalism entrepreneurs pay themselves for the risks they take while under pure socialism the state pays for everything, but a middle course is surely preferable; 'preferable' but also inevitable given that, almost by definition, capitalist and socialist regimes tend to be impure."[35] By that time Ernst was becoming well known for his skill at handling the dispute between modernism and traditionalism in architecture, an argument with which he would engage long after his forced emigration to London in 1933. He discarded decorative forms in favor of the building material's inherent qualities, and emphasized simplicity and clarity of form, open-plan interiors and absence of clutter. Much like the psychoanalysts for whom he built homes, Ernst called attention to the dialectical relationship between architect and client, construction and location, interior and exterior. Ludwig Scherk, a Berlin perfume manufacturer known as a patron of avant-garde and younger architects, commissioned Freud to build his private residence. So too did many psychoanalysts: Hans Lampl retained Ernst to build a house, and his clients eventually included Franz Alexander, Sandor Rado, Rene Spitz, Hugo Staub in Berlin, and Ernest Jones, Melanie Klein, David Matthew, Hilde Maas and Kaethe Misch Friedlaender, after 1933 in London. Where later modernist architects rejected tradition in order to discover radically new ways of making buildings, Ernst had fairly pragmatic views.

Instead of bluntly challenging traditional architectural norms, Ernst Freud questioned the character of a building and its suitability, and incorporated whatever pre-existing elements would agree with its new purpose: his signature intention to fuse modernist functionalism with transnational cosmopolitanism can be seen even at this early stage in his architectural career. Called on by Simmel to transform Schloss Tegel's mid-size castle and leafy suburban park reminiscent of pre-war Austria, Ernst remade the building into a working mental health facility in record time. He stripped several large rooms of their ornamentation and converted them into communal bathing and eating facilities that evoked the lines of Viennese *Gemeindebauten* buildings overseen by his former mentor, the urban architect Adolf Loos. Bauhaus-style white overhead lamps were hung from the ceiling to distribute light evenly over the patient and staff dining tables (the two groups shared communal meals). Hallways were cleared so that the rooms would open directly onto them, and a large area toward the back was fitted with an unusual round bathtub for hydrotherapy. The furniture was simple and bold, characteristic of Ernst's designs with deep-seated upholstered armchairs, round tables and the ubiquitous wood book-case. Clients, who paid little or nothing for their treatment, entered through a broad portico lightened by white stucco walls and walnut shutters opened to the garden, all of which conveyed an air of infor-mality rarely associated with a psychiatrist's office. For the Schloss Tegel analysts, many of whom were members of Simmel's Associa-tion of Socialist Physicians and who chafed at traditional approaches (either purely psychosomatic or purely emotional with little in between these extremes) to the treatment of psychiatric disorders, the design appealed both to their political progressiveness and to their clinical objectives.

Ernst Freud's modernist architectural ideas gained considerable transnational traction. Whether in the domestic or public housing spheres, he encouraged architects to develop new building forms and building methods and to join in the debates on town planning. "It is most surprising... that the whole idea of modern architecture has not begun to influence the features of English towns," he wrote later. He spoke of a new emphasis on the lawful voice of the building's dweller, much like the psychoanalyst legitimating the voice of the patient. "For the erection of modern buildings the existence of modern architects is not sufficient. Important above all are clients, inclined to accept and appreciate the principles of modern architecture."[36] His intention, for example, to contain all rooms of a house or a clinic within an almost square footprint, evoked the modern style of Loos's Austrian and Czech houses. Given the watered-down Expressionism popular in Berlin until well into the 1920s (also Art Deco, Socialist Realism, neo-classicism and its eventual National Socialist styles, and streamlined Moderne in var-ious places and at various times) Ernst Freud's Modernism made the question of architectural functionalism one of full accord between the individual and the environment. In *début de siècle* Vienna, investigat-ing human needs from the perspective of architecture was concerned with the alignment of private and public space, on the one hand the individual and family, and on the other hand, citizenship and the state.

This image of a prevailing symbiotic exchange between the individ-ual and the social environment allowed, perhaps for the first time, the prospect of counting all disenfranchised populations within the scope of government. Most significantly, this inclusion could, in theory, vir-tually eradicate "otherness" in feeling and in fact. The strength and durability of this idea, of the overriding impact of the environment on human development, made it ubiquitous in the construction of social institutions from housing to transportation, from family to religion. As they took up the task of rebuilding the psychoanalytic movement eroded by World War I, the psychoanalysts did so within this specific person-in-environment dialectical paradigm, itself a Social Democratic framework that would soon remake much of post-monarchy Europe. Psychoanalysis of the 1920s spread in the context of *début de siècle* Vienna's humanist and vigorous social welfare ideologies, and repositioned the analysts from their social margins (as intellectuals and Jews) to a new-found political nucleus actually predicated on "otherness." The real question behind the designation of "other" was "relative to which environment?" At this turning point in the history of the movement, the psychoanalysts pictured a humanist world governed by rational laws, laws that recognized the depth with which environments (family, home, clinic) impeded or advanced human progress. In the words of Siegfried Bernfeld, "Freud's psychoanalysis...originates – as do Marx and Engels – from love and hunger as the fundamental human drives. It investigates the influence of the environment on the basic drives of the individual."[37]

"Now that we have free clinics"

In the fertile first quarter of the twentieth century, at least eight Modernist movements articulated a transnational vocabulary designed in large part to integrate "environment" into art, architecture, design and literature: the British Arts and Crafts Movement (1880–1910), American Arts and Crafts (1900–1915), European Art Nouveau (1880–1905), American Art Nouveau (1890–1910), Wiener Werkstätte (1903–1933), De Stijl (1917–1928), Bauhaus (1919–1933) and European Art Deco (1920–1940). In some ways it was an exquisite exercise in unifying intellectual diversity and dispersion of the 1920s: who is the more revolutionary, Sigmund Freud or Alfred Loos? Psychoanalysis or architecture? Freud never said that his *theory* was specifically Marxist; it is only Marxist relative to the bourgeois institutions that demand conformity to pre-existing social norms and that, therefore, deny the impact of changing environments. So while Loos drew attention to all the accouterments of social democracy in his buildings and cafés, his theory of minimalism was actually bourgeois because it romanticized obsolete institutional standards. In contrast, Freud the rationalist, who never vied for the title of Marxist (and who utterly disavowed Communism), suggested a flexible *praxis*, a modernist response to human suffering; after all, everyone perceives the outer world subjectively through a dynamic web of drives and instincts. Just as cars had replaced horses, so psychoanalysis would do away with conventional bourgeois treatment inherited from the nineteenth century, and by opening community clinics, bring a transnational flow of help to people regardless of their outer worlds.

Once, the psychiatrist Joseph Wortis recounted, he and other American physicians went to Vienna to be analyzed by Freud, and as a skeptical young man he declared that Freud must surely know that "the special weapon of the positive transference... raises the whole question of the importance of money to patients in analysis." Perched on his seat in his consulting room, his eyes shining, Freud re-iterated his modernist adherence to universal access. "Now that we have free clinics and the psychoanalytic institutes... anybody can now be analyzed. They may have to wait a little, but everybody has the privilege." At one point, Wortis suggested that Freud seemed to think that the spread of free clinics would do little to make the world a better place. No, Freud answered confidently, "the present system seems best, and there is no occasion to worry about it."[38]

Freud's readiness to confront a bitter post-war society, and to redirect this acrimony into social activism through psychoanalysis, made the *début de siècle* analysts not just great theoreticians but also dedicated anti-traditionalists. Some have pushed the field further into transnational community-based practice; others have confined it to ever smaller institutional fragments. Even before he died, Freud saw psychoanalysis endangered by the lure of elitism and narcissistic medicalism. But nearly a century later, though Simmel's foresight and Jones's piety and Reich's politics seem to belong to an era less cynical than our own, Freud's uplifting modernism still has the power to transform, and to mobilize. The "conscience of society" has awakened.

Notes

1. Sigmund Freud, "Lines of Advance in Psychoanalytic Psychotherapy." In J. Strachey (ed. and trans.) *The Standard Edition of the Complete Psychological Works of Sigmund Freud* (Vol. 17), London: The Hogarth Press, 1918, p. 167.
2. Ibid.
3. Beth Noveck, "Hugo Bettauer and the Political Culture of the First Republic." *Contemporary Austrian Studies*, 3 (1995): 143; *Bettauer's Wossenschrift* (1924).
4. Richard F. Sterba, *Reminiscences of a Viennese Psychoanalyst*, Detroit: Wayne State University Press, 1982, p. 7.
5. Robert Coles, *Anna Freud: The Dream of Psychoanalysis*, Oxford and NY: Addison-Wesley Publishing Co., 1991, p. 113.

6. The term "modernism" is necessarily broad. Chronologically, the era of modernism begins in Europe and America around 1914 and ends with World War II, though some critics date it from the early 1890s (Harry Levin, "What Was Modernism?" *Refractions: Essays on Comparative Literature*, Oxford and NY: Oxford University Press, 1966, pp. 271–295). According to Philip Brookman's *Essential Modernism* (London: V & A Publications, 2007), the term denotes "the historic development of modern form, starting with the post–World War I philosophical and artistic search for Utopia. Its legacy [has] given us profoundly new ways of looking at the world." Also within the Utopian tradition, Timothy Reiss's *The Discourse of Modernism* (Ithaca: Cornell University Press, 1982) sees modernism as a meeting ground for scientific, political, linguistic and aesthetic theory from which a "new image of the rational self" emerges, leading to a new mode of discourse. It could well be said that psychoanalysis falls within this definition.

7. Michael Molnar (ed. and trans.), *The Diary of Sigmund Freud 1929–1939, A Record of the Final Decade*, New York: Charles Scribner's Sons, 1992, pp. 113–114.

8. Sigmund Freud to Sándor Ferenczi, Letter # 772, November 17, 1918 In Ernst Falzeder, and Eva Brabant, (eds). *The Correspondence* (Vol. 2), Cambridge, MA: The Belknap Press of Harvard University Press, p. 311.

9. Grete Bibring in Robert Coles, *Anna Freud: The Dream of Psychoanalysis*, Boston, MA: Addison-Wesley Publishing Co., 1991, pp. 110–115.

10. See Nathan G. Hale, *Freud and the Americans – The Beginnings of Psychoanalysis in the United States, 1876–1917*, Oxford and New York: Oxford University Press, 1971.

11. Wilhelm Reich, "The Living Productive Power, 'Work-Power,' of Karl Marx." In: Mary Boyd Higgins (ed.), *People in Trouble* Philip Schmitz (trans.) (1976) New York: Farrar, Strauss and Giroux, 1936, p. 75.

12. Sigmund Freud to Sándor Ferenczi, Letter # 762, October 11, 1918 . In Ernst Falzeder and Eva Brabant (eds) *The Correspondence* (Vol. 2) p. 299.

13. Sigmund Freud, "Lines of Advance in Psychoanalytic Psychotherapy." In J. Strachey (ed. and trans.) *The Standard Edition of the Complete Psychological Works of Sigmund Freud* (Vol. 17), London: The Hogarth Press, 1918, p. 167.

14. Ibid.

15. Ibid.

16. Paul Tillich cited in Peter Gay, *Weimar Culture – the Outsider as Insider.* New York: Harper and Row, 1968, p. 36.

17. Sigmund Freud, "Preface to Max Eitingon's 'Report on the Berlin Psycho-Analytic Policlinic'." In J. Strachey (ed. and trans.) *The Standard Edition of the Complete Psychological Works of Sigmund Freud* (Vol. 19), London: The Hogarth Press, 1922, p. 285.

18. The New York Psychoanalytic Society had just drafted their proposal for the New York Psychoanalytic Clinic, and agreed to petition the state for authorization. Presumably influenced by the increasingly conservative American Medical Association, the charter was denied by the NY State Board of Charities. For a comprehensive account, see Nathan G. Hale, *The Rise and Crisis of Psychoanalysis in the United States – Freud and the Americans, 1917–1985*, New York and Oxford: Oxford University Press, 1995.

19. Adolf Loos, "Ornament und Verbrechen." (1908) In Franz Glück (ed.) *Sämtliche Schriften*, Vienna, 1962, p. 280.

20. Else Pappenheim, "Politics and psychoanalysis in Vienna before 1938" Paper presented to the 1984 Oral History Workshop of the American Psychoanalytic Association.

21. Se Helmut Gruber, *Red Vienna – experiment in working-class culture, 1919–1934*. New York and Oxford: Oxford University Press, 1991, p. 51.

22. Sigmund Freud to Eitingon, October 25, 1918. Cited (by permission of Sigmund Freud Copyrights, Wivenhoe) in Peter Gay, *Freud, A Life for Our Time*, New York: Anchor Books, Doubleday, 1988.

23. Robert Coles, *Anna Freud: The Dream of Psychoanalysis*. Boston, MA: Addison-Wesley Publishing Co., 1991, pp. 110–115.

24. Else Pappenheim, transcribed and expanded "Remarks on Training at the Vienna Psychoanalytic Institute," Oral History Workshop of the American Psychoanalytic Association Meeting, December 1981, by permission.

25. Sigmund Freud to Franz Alexander, May 13, 1928 in letter to Franz Alexander, May 13, 1928. in Jones, Ernest. *The Life and Work of Sigmund Freud*, vol 3. New York: Basic Books, 1955, pp. 447–448.

26. Jones to Freud, #546, October14, 1929. In R.A. Paskauskas (ed.) *Sigmund Freud and Ernest Jones*, p. 665; also see Jones to Eitingon, October 18, 1929, document # CEC/F01/40, Archives of the British Psychoanalytical Society.

27. Sandor Rado's speech reported by the Frankfuster Zeitung of 27 February 1029, cited in Karen Brecht, Volker Friedrich, Ludger M. Hermanns, Isidor J. Kaminer and Dierk H. Juelich, eds. *"Here Life Goes On In A Most Peculiar Way..."* Psychoanalysis Before and After 1933, english edition prepared by Hella Ehlers and translated by Christina Trollope (Hamburg 1990), p. 57.

28. Otto Fenichel to Edith Jacobson, Annie Reich, Barbara Lantos, Edyth Gyomroi, George Gero, and Frances Deri, Runbrief #1, March 1934, Box 1, Folder 1, Austen Riggs Library.

29. Jones to Freud, Letter #476, February 25, 1926. In R.A. Paskauskas (ed.) *Correspondence*, p. 592.

30. See Russell Jacoby, *The Repression of Psychoanalysis: Otto Fenichel and the Political Freudians* (New York: Basic Books, 1983) for an excellent account of this history.

31. Benjamin to Brecht, cited from Benjamin, *GS VI*, p. 826, in Momme Brodersen, *Walter Benjamin – A Biography*, New York: Verso, 1996, p. 187.

32. Ernst Simmel, "Psycho-Analytic Treatment in a Sanatorium." *International Journal of Psycho-Analysis*, 10 (1929): 89.

33. Ernst Simmel, "Erstes Korreferat." In *Zur Psychoanalyse der Kriegs-Neurosen*. Doris Kaufmann, "Science as Cultural Practice: Psychiatry in the First World War and Weimer Germany." *Journal of Contemporary History*, 34/1 (1999): 140.

34. For a thorough exploration of Ernst Freud's oeuvre, see Volker M. Velter's, "Ernst L. Freud – Domestic Architect." In Shulamith Behr and Marian Malet (eds), *Arts in Exile in Britain 1933–1945 – Politics and Cultural Identity*, Amsterdam: Rodopi, 2005, pp. 201–240.

35. Gaetano Ciocca, Tempo 208 (May 20, 1943): 23–24, cited in Jeffrey T. Schnapp, *Building Fascism, Communism, Liberal Democracy – Gaetano Ciocca*,

Architect, Inventor, Farmer, Writer, Engineer, Stanford, CA: Stanford University Press, 2004, p. 245.

36. Ernst L. Freud (1924) *Design for Today,* cited in Volker M. Velter (2005) "Ernst L. Freud – Domestic Architect." In Shulamith Behr and Marian Malet (eds), *Arts in Exile in Britain 1933–1945 – Politics and Cultural Identity,* Amsterdam: Rodopi, 2005, pp. 201–240.

37. *Der Tag,* September 14, 1928, clipping in Folder "Professional File – Europe: 1858–1942," Container 5, papers of Siegfried Bernfeld, Collections of the Manuscript Division, U.S. Library of Congress.

38. Joseph Wortis, *Fragments of an Analysis with Freud,* New York and London: Jason Aronson, 1984, p. 151.

2
Beyond the Blues: Richard Wright, Psychoanalysis, and the Modern Idea of Culture

Eli Zaretsky

Introduction: the place of transnational psychoanalysis in African-American history

The transformation in the self-image of American blacks between 1903, when *The Souls of Black Folk* appeared, and the 1963 March on Washington is one of the most inspiring episodes in all of history. Early twentieth century black America was a ravaged continent, rural, proto-literate and impoverished, still dominated by lords and masters, whether plantation-owners and ex-confederates, internal despots such as Booker T. Washington, or Northern patrons, "philanthropists," and political mentors. A few years before *The Souls of Black Folk* appeared, when some black college students in Alabama accidentally wandered into a white railroad car, their black college president, William H. Council, called them into his office with the words, "You all have ruined me," and then, weeping, handed his resignation to the white board of trustees.[1] By way of contrast, similarly situated black college students launched the Civil Rights movement in the Deep South in the 1950s. Facing down intim-idation, jail, and lynching, their actions culminated in the March on Washington and the Civil Rights Acts of 1964 and 1965.

Needless to say, there is no simple explanation for this transforma-tion. World history, centered on the struggle against fascism, the overall democratization of America in regard to such matters as social class, ethnicity, and gender, and the unexpectedly potent cultural resources of black America in such areas as family structure and religion, all played important roles. The purpose of this essay is to call attention to one episode – Richard Wright's encounter with psychoanalysis in the Popular Front context of the 1930s and 1940s, as well in the

43

post-colonial aftermath of the Popular Front in the 1950s – in the inspiring achievement of black America in coming to grips with its past. The significance of this episode lies in the light it sheds on the central role of *trauma* in African-American history. The origins of black America lay in violent usurpation. Africans were taken from their homelands, separated from their families, kinship, and language groups, sold, resold, and sold again, kept in ignorance, beaten, raped, slaughtered at will, and, at best, condescended to or mocked. Even the abolition of slavery was more a matter of form – albeit an important form – than of content. Without question black Americans resisted these assaults. Still, any developments that could empower black America had to deal somehow with this intensely painful past. Any effort to remember, mourn, or work through would inevitably produce suffering, perhaps guilt, perhaps shame, and certainly rage. Psychoanalysis, I will argue, made a distinct contribution to this "working through" because it was, above all, a theory of trauma, by which it meant a violent rupture, wound, or tear, or a series of such wounds, which had overwhelmed the normal functioning of consciousness, thereby leaving an unavoidable unconscious conflict.

To appreciate the psychoanalytic contribution, one must begin with earlier efforts to deal with the black collective past. Although it certainly had many predecessors, the benchmark for these efforts is W.E.B. Du Bois's 1903 *The Souls of Black Folk*. Drawing on then current, mostly Hegelian but also social scientific, philosophies of culture, which defined human beings as symbol-producing animals, Du Bois proposed to counter racism by demonstrating black Americans' humanity. In particular, he showed that black Americans had maintained a culture under slavery and thus had a basis on which they could participate in American and world history. Needless to say, *The Souls of Black Folk* breathes with an intense, sometimes overwhelming awareness of black suffering. Complex and protean, the work is multi-layered. "The price of culture," Du Bois wrote presciently, "is a lie." Du Bois's thought also evolved after *The Souls of Black Folk*, for example, in debates with Alain Locke, Carl Van Vechten, and Jean Toomer. Nonetheless, the *core* approach taken in that work can be described as *affirmative* in the sense that its aim was to affirm black culture and thus black humanity, even under the most unfavorable conditions.

The affirmative approach that runs through *The Souls of Black Folk* was a familiar one in the early twentieth century. In 1903 many peoples, such as the Russians, Irish, Poles, Czechs, and Hindus were emphasizing both their cultural specificity and their overall contribution to human advancement as they became integrated into a global market and a global public sphere. Often, like Du Bois when he chose such terms as "soul" and "folk" they contrasted *Kultur*, the particularly German inflection of culture with its connotations of authenticity, depth, and inwardness to the artificiality and superficiality of *Civilization*. Such cultural self-definition was obviously a two-edged sword since it affirmed not only the aspiring nation, but also the dominant bourgeois narrative of progress and humanity. A necessary step in black America's recovery from slavery, this approach never fully satisfied most African-American thinkers, including Du Bois.

Whatever high value African-American thinkers placed on black resilience under slavery, they could never ignore the effects of racial violence. Over time, therefore, they developed an alternative view of culture, one that centered on the founding character of that violence and the traumatic dimensions of African-American identity. Paul Gilroy's *Black Atlantic* represents a recent example. As Gilroy pointed out, "the [slave's] repeated choice of death rather than bondage" (as typified by Margaret Garner, the slave who killed her infant daughter and who was the inspiration for Toni Morison's *Beloved*) articulated a principle of negativity which did not so much oppose as deepen and complicate the affirmative ideal of culture laid down by the dominant (white) culture and functional in the period of "modernization," i.e., adaptation to global capitalism, that characterized the early twentieth century.[2]

The limits of the affirmative approach can be grasped if we consider the blues. Originating in the late nineteenth century, the blues derived their power from the way in which they directly expressed the pain, the anguish, and the losses of black life. Thus Blind Willie Johnson moans: "My mother is dead"; Ma Rainey laments her betrayal by her man, and Fred MacDowell describes himself as lost, humiliated, and unmanned. Because of the emphasis on loss, the blues seems miles away from the affirmative approach. And yet there remains an affinity between them: just as Du Bois affirmed black humanity's cultural self-expression under slavery, so the blues affirmed the triumph of the singer over his sorrow, that is of art over reality. Reflecting the still closely felt presence of a great wrong or injury, moreover, blues singers returned obsessively to their pain, but offered no path beyond it.

The project of moving beyond this impasse – beyond the blues – can serve as a synecdoche of black empowerment. That project had a natural affinity with psychoanalysis. The discovery of the unconscious, like the full discovery of the African-American past, was a discovery of the "dark side" of modernity, the underside that had been disowned by the dominant narrators and had become all the more powerful as a result. In trying to bring this traumatic underside into consciousness, African-American artists and intellectuals experienced similar difficulties – similar impasses – to those psychoanalysts encountered when working with individuals. Beginning by translating the cultural unconscious directly into consciousness – that is by recounting the racial past in a way that presumed the rationality of the narrator and the audience – they encountered *cultural defenses* or *resistances*, in part reflecting the obdurate intensity of racism, and in part reflecting the shame, guilt, and anger of the African-American community itself. They learned, accordingly, that the road to the past could never be direct but had to proceed *through* the resistance, that is through shame, guilt, and anger. This approach, which I shall call *resistance*, deepened and transformed affirmation. It might also be called deconstructive by analogy to Jacques Derrida's view that the meaning of a text can never be understood directly but only through working through the text's defensive operations. But whether it is termed resistant or deconstructive, this approach holds that it is not the *trauma* that needs to be resisted or deconstructed, it is the *affirmation*; the trauma needs to be accepted.

My argument unfolds in three steps. In the first step I describe the discovery and first formulation of the problem of a traumatized past in the work of W.E.B. Du Bois and the Harlem Renaissance. Notwithstanding the prior contribution of slave narratives, *The Souls of Black Folk* largely invented the idea of black culture as a resource for the development of the race, an invention that by politicizing African-American cultural thought set it on its distinctive path. Although Du Bois's emphasis on culture had an affirmative purpose, it also necessarily addressed the problem of working through the trauma and violence of the black past. As many passages in *The Souls of Black Folk* suggest, Du Bois intuited that this working through is a *discursive* project. That is, the artist or cultural leader plays the role of a *witness*: someone who attests to the crimes that have been committed, someone who themselves has suffered what his or her people have suffered, and someone who gives voice to those who have suffered passively, and, as a result, have lost their own capacity to testify. Du Bois grasped, then, that the black artist was in an antiphonal call and response relationship to his own people, and not merely to cultural arbiters and that this relationship had something to do with the void of language left in the wake of trauma.

Nevertheless, as Du Bois himself later remarked, the key concepts of *The Souls of Black Folk*, such as invisibility, double-ness and the "veil," were still "pre-Freudian," in the sense that they did not place violence, trauma, and disruption at the center of their world. Such Harlem Renaissance figures as Jean Toomer and Zora Neale Hurston grappled with this limitation. Building on Du Bois's work, they insisted that Africa's disastrous encounter with Europe, the middle passage, slavery, pseudo-emancipation, lynching, and Jim Crow had initiated "a dramatic loss of identity and meaning, a tear in the social fabric" of African-American culture, a loss that was "played over and over again in individual consciousness."[3] Feeling that black America, only recently rescued from the ravages of slavery, struggling to find a footing in the new mass, industrial society, needed to draw upon all of what Tom Nairn has called its "inherited and... unconscious powers" to confront an "inescapable challenge," that of creating its own culture, they hoped to enlist psychoanalysis in that project.[4] Thus, their work – although inevitably affirmative – also pointed beyond affirmation in principle, if not yet fully in fact.

In a second step, I locate the fuller shift from affirmation to resistance in the life and work of Richard Wright. If Du Bois stressed continuity and progress, Wright emphasized lack, discontinuity, and failure. If Du Bois, following Hegel, described culture as *self*-discovery, the cultivation of inner as well as outer resources, Wright stressed trauma, division, and rift. If Du Bois viewed "doubleness" intersubjectively, in terms of the relationship between white and black, Wright wrote works of mourning and "working-through." As we shall see, Wright's appropriation of Freud was inseparable from his involvement in the Popular Front. Drawing on the Marxist milieu that surrounded him, Wright developed an agonistic view of culture, as rooted in trauma and oppression. But while the agonism came from Marx, Wright added a distinctly non-Marxist focus on individual subjectivity. In developing his views of the psychology of oppression and rage, Wright became the first African-American intellectual to fully engage with Freud. His highly personal encounters with psychoanalysis, motivated by his intense refusal *not* to be defined – as a black Southerner, as a Communist, and as an American – accompanied the larger shift in psychoanalysis from a theory of the unconscious to a theory of the resistance.

Part Three describes Wright's efforts to consolidate the introspective gains achieved by his confrontation with racial violence. Although Wright left the Communist Party in 1942, his thinking remained centered on the "uncanny" resemblance of fascism to slavery. Attempting to work through his core obsession, the hunt, capture, and lynching of slaves and their successors, Wright moved to France in 1947. There he elaborated his Popular Front insights to describe a world in which outsiders were at the center while the dominant classes were periph-eralized. His later work foretold a paradigm shift that recognized and legitimated the psychological and cultural place of violence in the pro-cess of emancipation. Recognized by both Jean Paul Sartre and Frantz Fanon, and developed, albeit critically, in the work of such contempo-rary thinkers as Harold Cruse and Paul Gilroy, this shift is distinguished by the idea that ethnocide is not a violation of the emancipatory project of modernity, but its defining characteristic.

W.E.B. Du Bois and the Harlem Renaissance

W.E.B. Du Bois established the basic premise of all African-American cultural thought: the need for a collective project centered on the recon-struction of black culture. A founder of the National Association for the Advancement of Colored People (NAACP) who appreciated and empha-sized the struggle for individual rights, Du Bois nevertheless followed Hegel in arguing, "consciousness is not achieved individually and one by one, but rather through the people, each people rising to a con-sciousness of itself."[5] While, in general terms, this view remained indis-pensable, it had three limitations that later thinkers, including Du Bois, sought to address. First, it was based on an organic, holistic view of cul-ture, one that ignored the significance of traumatic disruption. Second, it assumed a resonance or even harmony between the individual and the group, rather than stressing their diremption and discontinuity. Finally, it stressed the national as opposed to trans-national or even global con-text. The genealogy of the term "culture" illuminates these problems.

When the term culture emerged in the Middle Ages it meant tillage, husbandry, and the cultivation or rearing of a plant or crop; it included therefore, from the first, the implication of organic or natural develop-ment. During the nineteenth century – in other words, with the first industrial revolution – romantic poets and philosophers extended the term to refer to separate "folks," peoples, nations, or "races," each, as Johann Herder famously argued, with its own spirit, language, or "cul-ture." The evolution of culture – which Hegel called "objective spirit" – was thought to coincide with the evolution of the individuals within culture. The romantic poets, according to Raymond Williams, "discov-ered" English culture, in the sense of a communal reserve that could be used to withstand the rationalizing tendencies of industrial capital-ism. Still later, as Tom Nairn argued, as different peoples entered into the global economy, they found that their main resource was their col-lective selves – not just their labor, but their inherited resources, their linguistic, or musical gifts, their writers, crafts, sacred places, and texts. When anthropologists, such as Franz Boas, began to use the term culture to refer to the humanly created, as opposed to biologically given, elements of human society, this gave the term special relevance to the struggle against racism.

The Souls of Black Folk exemplifies this tradition. As is well known, the work was a riposte to Booker T. Washington's 1901 *Up from Slavery*. While Washington was concerned with training the black masses to take their place in industry and commercial agriculture, Du Bois wanted to encourage a black middle class, the so-called talented tenth, defined in good part by its cultural achievements but also able to inspire the race as a whole. In pursuit of this goal, *The Souls of Black Folk* emphasized the need to engage with the past, beginning with the riches of Africa itself.

Part of the strange magic – the modernity – of *Souls of Black Folk* lay in Du Bois's understanding of the difficulty involved in that task. Writing of the fragments of music that he placed at the start of each chapter, Du Bois recounted: "Ever since I was a child these songs have stirred me strangely. They came out of the South unknown to me, one by one, and yet *at once I knew them as of me and mine.*" They descended, he explained, from his "grandfather's grandmother seized by an evil Dutch trader two centuries ago," a woman who "crooned a heathen melody to the child between her knees." Passed down through the generations, the melody reached Du Bois in the following form:

> *Do bana coba, gene me, gene me!*
> *Do bana coba, gene me, gene me!*
> *Ben d'nuli, nuli, nuli, nuli, ben d'le.*

Recognizing the precious quality of such fragments, Du Bois stressed their role in inspiring black leaders. Later he confessed he had no idea where the melody came from.

Given Du Bois's understanding of the complications involved in the past, what could psychoanalysis have added? Consider the most famous passage in *The Souls of Black Folk*:

The Negro is a sort of seventh son, born with a veil, with second-sight in this American world, – a world which yields him no true self-consciousness, but only lets him see himself through the revelation of the other world. It is a peculiar sensation, this sense of always looking at one's self through the eyes of others, of measuring one's soul by the tape of a world that looks on in amused contempt and pity. One ever feels his two-ness, – an American, and a Negro; two souls, two thoughts, two unreconciled strivings; two warring ideals in one dark body, whose dogged strength alone keeps it from being torn asunder.

This passage laid out the problems that would occupy subsequent generations of black intellectuals: the problem of being defined by the gaze of the other, of having no true self-consciousness, of being internally divided, of being excluded from the human community, that is from the modern idea of culture. Nevertheless, Du Bois's "doubled consciousness" had nothing to do with the unconscious in the Freudian sense. As with all of Du Bois's thinking at that point, its roots lay in German idealism, and specifically in Emersonian transcendentalism, which stressed the inevitable gap between empirical reality and the transcendent realm, a gap that Du Bois's teacher, William James, applied to the study of multiple personality. Du Bois's "doubleness" was interpersonal, sociological, and phenomenological. The "dialectical overcoming" it aspired to has been well summarized by Paul Anderson:

First, the cancellation of racial hierarchy, its alienating consequences...second, the preservation of certain essential cultural differences; and, finally, the elevation of a hybridized African American identity.[6]

As Du Bois explained, the goal was "to make it possible for a man to be both a Negro and an American." Certainly this goal, intrinsic to the modern idea of culture, was and is indispensable. But it was also insufficient. The psychoanalytic version of "doubleness," which would describe blacks not as "divided" but rather as having the largest part of their collective past unavailable to consciousness, and all the more potent because it was unavailable – this concept of the unconscious was missing in Du Bois.

The first significant interaction between African-American intellectuals and psychoanalysis occurred during the Harlem Renaissance when such figures as Jean Toomer and Zora Neale Hurston referred to the "racial unconscious." For Toomer and Hurston, the racial unconscious was a *collective* unconscious or rather, in Freudian terms, *preconscious*, meaning sensually charged material (words, speech patterns, images, memories, musical rhythms, gestures) that are not normally part of conscious awareness, but are available to consciousness; that is, they are not repressed. Proto-Freudian at best, the significance of this concept of the racial unconscious lay in the intuition that culture was disrupted by trauma and discord.

Hurston's ethnographic journeys into the American South exemplify the meaning of the "racial unconscious." A student of Boas and of Edward Sapir, the latter being the main figure responsible for introducing psychoanalysis into American anthropology, Hurston went South in search of African-American "folk" culture. At first she complained that there was no such thing since Southern blacks were interested only in radio and the movies. Later she discovered a folk culture in what Frances Lee Utley called "the arts and crafts, the beliefs and customs of our lumber camps, city evangelical storefront churches, back-alley dives, farmer's festivals and fairs, hill frolics, carnivals, firemen's lofts, sailor's cabins, chain gangs, and penitentiaries."[7] According to Hurston, the black South's dialect, tales, humor, and folk mores constituted a collective, aural catalogue of the past, a past that was still insistently present, and that had its own character. Thus, "the white man thinks in a written language and the Negro thinks in hieroglyphics."

The analytic sensibility came out in Hurston's stress on dissonance and negativity rather than organic holism. Directly referencing Ruth Benedict's *Patterns of Culture* which argued that cultures had a kind of unifying style or pattern, manifest not only in conscious, intentional activities such as architecture or literature, but also in unexpressed assumptions or forms, Hurston's "Characteristics of Negro Expression" described lack of reverence, "angularity," redundancy, mimicry, and, especially, "restrained ferocity in everything," as the axes of black America's unconscious grammar.[8] As with Du Bois, behind Hurston's affirmative approach lay sensitivity to the traumatic past of the African-American people. Her best-known work, *Their Eyes Were Watching God*, traces the efforts of a black woman to work through her own traumatic experiences by drawing upon the language, folk patterns, and collective memories of the Deep South. It was only when Hurston's heroine turned her culture – the racial unconscious – into something autonomous, personal, and idiosyncratically hers that she became what Henry Louis Gates later called "a speaking black subject."[9]

Jean Toomer, the author of *Cane*, also linked the idea of a racial unconscious to a disrupted, fragmented, painful cultural past. In a 1921 review of Eugene O'Neill's *The Emperor Jones* he wrote: "the contents of the unconscious not only vary with individuals; they are differentiated because of race...Jones lived through sections of an unconscious which is peculiar to the Negro. Slave ships, whipping posts, and so on...[H]is fear becomes a Negro's fear, recognizably different from a similar emotion, modified by other racial experience." *The Emperor Jones*, Toomer concluded, is "a section of Negro psychology presented in significant dramatic form."[10] Dissonance, for Toomer, also accompanied the transition to modernity. Thus, as Werner Sollors has written, *Cane*, with its strong visual images and musical effects was "an attempt at finding a literary equivalent for the dislocations that modernity had wrought by moving people from soil to pavements, making them ashamed of their traditional folk culture or changing it into commercial entertainment."[11]

The fullest appropriation of psychoanalysis by a Harlem Renaissance author, Hurston's *Moses, Man of the Mountain* (1939) introduced two further elements. First, Hurston used Freud's rewriting of the book of Exodus to stress the disrupted, impure, or "mongrel" character of African-American identity. Inspired by Freud's claim that Moses was both Egyptian and Hebrew, she identified African-Americans with Hebrews, the "chosen people," "de people dat is born of God," while also valorizing "passing," that is, the liminal, non-essentialist, and even performative character of identity. Secondly, Hurston emphasized the persistent costs of slavery to the struggle for freedom. Likening the long, slow process of African-American emancipation to the Hebrew's long period of wandering in the desert, she insisted that the people born in slavery have to die out before a new generation can gain emancipation. Like Wright, too, she intuited that freedom might pose greater challenges than slavery to the African-American people. Freedom, her Moses repeatedly tells his people after they have fled Egypt, is an inner state, "not a barbecue."[12]

With all of its enormous richness and complexity, the overall aim of the Harlem Renaissance was to bring out the humanity of the American Negro past, the sense in which black lives had not been negated by slavery, not been reduced to mere haulers of wood, tobacco, and cotton, to what Aristotle called *zoon*, mere animal existence. What Du Bois, Toomer, Hurston, and many others showed was that even under the horrendous conditions of slavery, blacks had retained their music, history, humor, folk tales, sexual practices, religion, and family ties, in a word their *culture*. To be sure, Du Bois, Toomer, Hurston, and others described African-American consciousness as divided, conflictual, dual: "two warring ideals in one dark body." But the reason for this duality was that African-American society had a telos, that of freedom, which still remained to be realized. *Soul*, as defined by Du Bois, was the acceptance of this ambiguous, in-between state, an acceptance based upon the unjust, sad (blue) but ultimately comic, in the sense of triumphant, direction of the black telos. The humor and sensuality with which Hurston leavened bitterness, the irony with which Toomer portrayed black strivings for upward mobility, the enormous spiritual resources in Du Bois's oeuvre, the blues: these are all examples of "soul," that is of the spirit's triumph over adversity. The bittersweet quality of the triumph is what gives black literature, music, and visual art its deeply moving quality.

As I noted, the blues became central to the modernist moment in African-American history because it expressed the individual's pain at being trapped between telos and reality. Although descended from collective sources such as the spirituals, work songs, protest songs, and field hollers, the blues was the first explicitly *personal* form in the history of African-American music. Representing a great cultural milestone in African-American cultural history, the emancipation of the individual from the group, the power of the blues nevertheless derived from its resonance with collective experiences of impasse, passivity, and stasis. The traumatic origins of African-American society made themselves felt in the anguish of the singer, the mournful humming that so often accompanied the guitar, and the prevalence of such themes as blindness, old age, and impotence. As Ralph Ellison explained:

> The blues [was] an impulse to keep the painful details and episodes of a brutal experience alive in one's aching consciousness, to finger its jagged grain, and to transcend it, not by the consolation of philosophy but by squeezing from it a near tragic, near comic lyricism. As a form, the blues [was] an autobiographical chronicle of personal catastrophe expressed lyrically.[13]

The African-American project of cultural reconstruction could not and did not rest content with the approach of the blues. The core problem was that this approach provided only sporadic access to the emotion that Aristotle claimed was closest to reason, namely anger. It was only in the 1930s, with the discovery of a great evil at the core of Western civilization, an evil that was racist in its very heart, namely Nazism, that African-Americans truly began to seek a pathway "beyond the blues." Du Bois, who had moved toward communism ever since visiting the Soviet Union in 1927, intuited the relevance of psychoanalysis to the departure that fascism encouraged in Western historical and cultural thought. As a young man, he recalled in his 1940 autobiography *Dusk of Dawn*, he thought of the "Negro problem as a matter of systematic investigation and intelligent understanding. The world was thinking wrong about race, because it did not know. The ultimate evil was stupidity. The cure for it was knowledge based on scientific investigation." Then "there cut across this plan which I had as a scientist, a red ray which could not be ignored...a poor Negro in central Georgia, Sam Hose, had killed his landlord's wife. I wrote out a careful and reasoned statement concerning the evident facts and started down to the *Atlanta Constitution* office.... I did not get there. On the way news met me: San Hose had been lynched, and they said that his knuckles were on exhibition at a grocery store...I turned back to the university, I began to turn aside from my work." As Du Bois later concluded: "in the fight against race prejudice, we were not facing simply the rational, conscious determination of white folk to oppress us; we were facing age-long complexes stuck now largely to unconscious habit and irrational urge." Criticizing his earlier view that "race prejudice [was] based on wide-spread ignorance," a view that buttressed the Hegelian goal of recognition and dialectical overcoming, Du Bois concluded that he had not been "sufficiently Freudian to understand how little human action is based on reason."[14] Du Bois, in this passage, recognized that there was a defensive trend in his earlier understanding. Alternatives to this trend emerged especially in the 1930s. We will focus on one: the life and writings of Richard Wright.

Richard Wright and the Popular Front

Writing against the backdrop of a traumatized past black intellectuals forged a powerful sense of the cultural past in the hope of empowering individuals. With Richard Wright, the project was reversed. Wright began with the individual, indeed, with that paradigmatic psychoanalytic concept, resistance, in the belief that individual self-assertion transcends social conditions. For Wright, as Edward Margolies has written,

> the achievement of freedom is an individual and private affair, a struggle that is won in the hearts of some men regardless of the external and environmental circumstances in which they find themselves. Their freedom is their rationality, but they discover themselves outsiders... to the myth bound passions and irrationalities of modern civilizations. Finally, they attempt to change or alter their environment so that it will conform to the freedom they know in their hearts.[15]

Ironically, what gave Wright the courage to wage his deeply personal struggle was communism.

Wright was born in 1908 on a plantation near Natchez, Mississippi. His father was a sharecropper who abandoned the family; his mother and her female relatives were religious zealots. To understand the quasi-totalitarian segregationist environment of his childhood, it is helpful to begin with Orlando Paterson's notion of slavery as social death. Slavery, Paterson stressed, was the result of defeat on the field of battle. The slave was incorporated into the society as an internal enemy, a non-being; violent death was not avoided but postponed in the form of a threat. Thereafter the slave had no social existence except as mediated by the master; since social death was a ritual that required constant reenactment, powerlessness, and dishonor constituted the main experiences of life. Later Wright wrote of his childhood: "I had already grown to feel that there existed men against whom I was powerless, men who would violate my life at will" "If I was a nigger," a coworker later told him, "I'd kill myself." How would the child of a sharecropper, the grandchild of slaves, respond to this condition?[16]

Wright responded by consciously negating every message that society, including his family, directed at him. In the words of Abdul R. Janmohamed, Wright established "a specular relation with society's attempt to negate him," meaning that Wright internally negated every negation, thus affirming himself. "In what other way," Wright asked in his autobiography, *Black Boy*, "had the South allowed me to be natural, to be real, to be myself except in rejection, rebellion, and aggression?"[17] Given this limitation, it's no wonder that Wright believed that freedom began with an *inward* struggle. Wright's rebellion, moreover, may have been invisible in the sense of being internal or psychological, at least at first, but it was not blind; it had form based on an awareness of the ritualized character of social death, and of the constant appeal of the gamble, the throw of the dice, the provocation of fate in a moment of violence in which the gods are forced to make a decision for or against the individual. Thus, Wright wrote of his hero in *The Long Dream*:

He knew deep in his heart that there would be no peace in his blood until he had defiantly violated the line that the white world had dared him to cross under the threat of death.... A mandate more powerful than his conscious will was luring him on, subsuming the deepest layers of his being.[18]

In 1925, at the age of 17, Wright left Mississippi for Memphis and then, two years later, went to Chicago. Responsible for the support of his sick mother, Wright found his early experiences of intimate, ritualized violence reenacted in the city. According to Janmohamed, "His inability to prevent his resentment from registering on his face or in his demeanor result[ed] in his dismissal from various jobs because his employers [did] not like his 'looks.' " On one occasion, Wright was fired for saying "yes sir, I understand," since that showed more self-respect than a Southern black was meant to possess. According to Wright: "I could not make subservience an *automatic* part of my behavior. I had to feel and think out each tiny item of racial experience in light of the race problem, and to each item I brought the whole of my life." Literature alone, he says, allowed him to stay "alive in a negatively vital way." Above all, H.L. Mencken inspired him. As he wrote in *Black Boy*: "this man was fighting, fighting with words. He was using words as a weapon, using them as one would use a club."[19]

In his Depression-era jobs Wright encountered the problem that Ellison later termed invisibility. As a young hotel bellboy he was summoned to rooms where naked white prostitutes lolled around as if he wasn't there. As Wright observed: "blacks were not considered human beings anyway...I was a non-man, something that knew vaguely that it was human but felt that it was not...I felt doubly cast out."[20] Wright also worked as an orderly in a Chicago medical research institute, an experience that underlined the significance of voice or language to the working through of trauma:

> Each Saturday morning I assisted a young Jewish doctor in slitting the vocal cords of a fresh batch of dogs from the city pound....I held each dog as the doctor injected Nembutal into its veins to make it unconscious; then I held the dog's jaws open as a doctor inserted the scalpel and severed the vocal cords. Later, when the dogs came to, they would lift their heads to the ceiling and gape in a soundless wail. The sight became lodged in my imagination as a symbol of silent suffering.[21]

The Communist Party, which Wright entered in 1932, gave Wright the intellectual, historical, and cultural grounds to sustain his negativity by demonstrating to him that the modern social order, beginning with what Marx called "primitive accumulation," was the product of wanton expropriation, murder, and theft. The product of the global transformation wrought by World War I, the Party was internationalist, proletarian, and oriented to the role of action and will. In *Native Son* Wright wrote, there is "no agency in the world so capable of making men feel the earth and the people upon it as the Communist party."[22] And in *The*

God that Failed he credited the party with giving him the "first sustained relationships of my life."[23] Marxism, he added, gave him his "first-full-bodied vision of Negro life in America," taking in and accommodating for the great migration as a whole, north and south, rural and urban, male and female.[24]

The appeal of Communism to African-Americans – Claude McKay, Countee Cullen, Langston Hughes, Alain Locke, Paul Robeson, and W.E.B. Du Bois were only a few of the figures active in or around the Communist party – had begun during the huge upsurge in black strikes or strikes involving blacks during World War I. At first, however, what attracted Wright to the Party was not its critique of capitalism but its contribution to the project of national cultural revival. Ever since the Russian Revolution, the "labor question" had been largely subsumed into the "national question," and in the 1930s the Communist International described the African-American "nation" as the vanguard of "colored peoples" throughout the world. For a while Wright's favorite book was Stalin's *Marxism and the National and Colonial Question.* "Of all the developments in the Soviet Union," he later recalled,

> the way scores of backward peoples had been led to unity on a national scale was what had enthralled me. I had read with awe how the Communists had sent phonetic experts into the vast regions of Russia...I had made the first total emotional commitment of my life when I read how the phonetic experts had given these tongueless people, a language, newspapers, institutions. I had read how these forgotten folk had been encouraged to keep their old cultures, to see in their ancient customs meanings and satisfactions as deep as those contained in supposedly superior ways of living.[25]

Soviet Communism's project of retrieval and affirmation continued and built upon the modern idea of culture that had inspired Du Bois. But it also added something new: a justification for that moment of irreconcilable negativity – the moment of violence – that Wright believed individual self-respect necessitated. Wright's "Long, Black Song," written in the late thirties, exemplifies what Communism added to the Harlem Renaissance. The story concerned a white traveling salesman, selling phonographs with clocks (the symbol of capitalist time discipline), who seduces a black woman, Sarah, by playing records of spirituals for her. When Sarah's husband, Silas, learns what happened he beats his wife, kills the salesman, and waits for the lynch party so he can kill a few more of his oppressors. Wright portrays Silas as a quintessential bourgeois who realizes the inadequacy of bourgeois values when all he owns can be so easily taken from him.

In "Long, Black Song" the spirituals are no longer the resource for the race which they were in *The Souls of Black Folk*. Rather, they have been co-opted into a crass, commercial culture that rests on racism and violence; like Sarah, they have been raped. This new understanding of the limits of the affirmative approach to culture helps explain Wright's critique of the Harlem Renaissance. Condemning the "parasitic and mannered" writers of the 1920s who, Wright claimed, "entered the Court of American Public Opinion dressed in the knee-pants of servility, curtsying to show that the Negro was not inferior, that he was human, and that he had a life comparable to that of other people," Wright singled out Hurston for special condemnation. Conceding that she "can write," he complained that her prose was nonetheless "cloaked in that facile sensuality... that whites like to associate with Negroes." "We are not attempting to restage the 'revolt' and 'renaissance' which grew unsteadily and upon false foundations ten years ago," he concluded.[26]

Wright needed to gain access to negativity and resistance only partly because the conditions of black life were so overwhelmingly oppressive. As a writer, his fundamental problem was that his subjects were non-introspective. Relying on onomatopoeia – rifles that CRACK!, whips that "whick," steam that goes "Psseeezzzzzzzzzzz" – foregrounding lynching, rape, murder, and the fugitive's futile escape, pressuring, as Peter Brooks has written, "the surface of reality (the surface of the text) in order to make it yield the full, true terms of his story," Wright's characters gain access to subjective self-awareness only when they realize how deeply they hate, invariably as the result of committing an act of violence. In the end, however, self-knowledge gained in this way proves futile. Thus, in *Native Son*, Bigger Thomas smothers his drunken, liberal, communist, female employer because he fears being discovered alone with her. "Having been thrown by an accidental murder into a position where he had sensed a possible order and meaning in his relations with the people about him," Wright wrote of Thomas, "having accepted the moral guilt and responsibility for that murder because it made him feel free for the first time in his life; having felt in his heart some obscure need to be at home with people and having demanded ransom money to enable him to do it – having done all this and failed, he chose not to struggle any more."[27]

Through Communism, then, Wright came to believe that the African-American struggle for equality could never be based on the demand for recognition alone, but had to give voice to the subjectivity of would-be violence. In that sense, Communism prepared him for his later interest not only in Sartre, the philosopher who preached the *agonistic* character of recognition but also, as we shall see, in Freud, whom Wright had been reading throughout the thirties, as we know from his library and from interviews.[28] As *Native Son* suggested, Wright regarded capitalism more through the lens of existential alienation than that of economic exploitation. In *The Outsider*, the main character explained: what "makes one man a Fascist and another a Communist might be found in the degree to which they are integrated with their culture. The more alienated a man is, the more he'd lean toward Communism." Communism, Wright added, was "something more recondite than mere political strategy... it was a *life* strategy using political methods as its tools... Its essence was a voluptuous, a deep-going sensuality that took cognizance of fundamental human needs and the answers to those needs,... it was a non-economic conception of existence."

The Popular Front, then, both as an experience and as an idea, helped move Wright beyond the idealist conceptions of culture that characterized early black cultural history. As an experience, the Popular Front sanctioned Wright's resentment and fury, practically, by encouraging his writing, emotionally, by its support for the African-American struggle for recognition and justice, and intellectually, by arguing that violent appropriation, as in Marx, not the dialectical unfolding of objective spirit, as in Hegel, determined the course of modern history. But there was something else about the Popular Front, something that opened the way for Wright's deepening understanding of the traumatic sources of African-American and, indeed, all Western history. For the *idea* of the Popular Front defined itself *negatively*, that is in terms of what it was *not*, what it was *against*, namely fascism.

It is probably impossible to overestimate the significance of fascism as a background condition for the intellectual and cultural revolutions of the World War II period. With its bases in organized, communal, ritualized, biologistically conceived violence, its obsessions with the body, disease, and death, its physiocratic cartographies of race, its anti-Semitism, fascism had a two-sided character. On the one hand, it was ultra-modern, utilizing the mobilization, high-tech planning and propaganda characteristic of the second industrial revolution. On the other hand, it was not only regressive but, to use Freud's term, uncanny. To someone with Wright's sensibility, it must have appeared as the return of something familiar but repressed: namely, a new edition, so to speak, of the infancy of Western civilization itself, its violent ethnocides, the hunt of slaves, the plantation *lager*, the mass transport of individuals, their separation into categories according to age, sex or nationality, their use as beasts of burden, the stacking of bodies in the holds of ships, the careful, clerically sanctioned accounting procedures, the legitimating role of money and bureaucratic rules, the ubiquity of death, cruelty and violence, everything that one could glimpse with terror from the corner of the nursery. Fascism was no caesura; it differed importantly from the nationalist mobilizations of World War I; it was not a response to Communism, nor an expression of the crisis of "monopoly capital," as the Communists sometimes suggested. It was a return of the terrible scenes from the infant years of Western expansion, scenes of death, castration, and torture that constituted the bad faith of Western liberalism, the "other" that the modern idea of culture claimed to *Aufheben*.

The fact of fascism, and its framing by the Popular Front, opened the way for a rethinking of Western liberalism, but it also helps explain the appeal of psychoanalysis. While the place occupied by psychoanalysis in culture of the 1920s, and again, in the culture of the late 1940s and 1950s, is well known, much less has been written on its place in the era of the Great Depression and World War II. In spite of such predictable magazine articles of the Depression as "Farewell to Freud," and "The Twilight of Psychoanalysis," and in spite of the equally predictable Communist anti-Freudianism, psychoanalysis flourished in the 1930s and 1940s in both England and America.[29] Part of the background common-sense of the Popular Front mentality, psychoanalysis was on a position to influence Wright's developing conception of culture, introducing traumatic disruption into organic wholeness, stressing the discontinuity between the individual and the group, and helping to shift the terrain from the African-American to the trans-national and global context. The uncanny resemblances of slavery and fascism, as well as the Popular Front's legitimation of violence, encouraged Wright's rethinking of the project of cultural reconstruction, so that resistance, the dawning awareness and acceptance of trauma, came increasingly to the fore.

The beginnings of this change can be observed in two autobiographical works: "How Bigger was Born," and *Black Boy (American Hunger)*, which Wright wrote after finishing *Native Son*. One of Wright's most powerful childhood memories, in which he recalled a friend, Carlotta, anticipate these works: "One day I stood near her on the school ground; we were talking and I was happy. A strong wind blew and lifted the black curls of her wavy hair and revealed...a long, ugly scar." The rawness and violence of the sight never left him.[30] This memory tied traumatic violence to women and to sexuality, a theme to which we will return. Meanwhile, *Black Boy* was explicit on the traumatic costs of slavery to African-American history: "After I had outlived the shocks of childhood, after the habit of reflection had been born in me, I used to mull over the strange absence of real kindness in Negroes, how unstable was our tenderness, how lacking in genuine passion we were, how void of great hope, how timid our joy, how bare our traditions, how hollow our memories, how lacking we were in those intangible sentiments that bind man to man, and how shallow was even our despair. After I had learned other ways of life I used to brood upon the unconscious irony of those who felt that Negroes led so passional an existence! I saw that what had been taken for our emotional strength was our negative confusions, our flights, our fears, our frenzy under pressure." This passage's painful acknowledgment of the continuing effects of the racial past has few precedents in African-American history.

In the course of the Popular Front, then, the literally unbelievable enormity of the crimes that had been committed against African-Americans began to rise to the level of true consciousness, and men and women began to bear the anguish that the realization of a traumatic past entails: the discovery that one's life could have been entirely different, that one's best potential had been destroyed by the rapacity of others, that the sanctity of one's heart had been cruelly violated by the instruments of an enemy, that one's parents, grandparents, and ancestors had had their lives stolen from them too. This was a moment that passed beyond the blues.

Ralph Ellison's review situated *Black Boy* in that context. In its refusal to offer solutions, Ellison conceded, *Black Boy* shared something with the blues since the blues "offer no scapegoat but the self." But aside from that, *Black Boy* was crucially different. Because Wright so directly faced the element of defiance and existential rage that came out of the African-American past, and was able to express the shame, sadness, and regret that accompany any genuine engagement with trauma, *Black Boy* transcended the blues: "In it thousands of Negroes will for the first time see their destiny in public print...In this lies Wright's most important achievement: he has converted the American Negro impulse toward self-annihilation and 'going under-ground' into a will to confront the world, [and] to evaluate his experience honestly."[31] In his review of *Native Son*, Irving Howe made much the same point. *Native Son*, he wrote, "made impossible a repetition of the old lies.... [It] assaulted the most cherished of American vanities; the hope that the accumulated injustice of the past would bring with it no lasting penalties."[32] Leaving the party in 1942, Wright began an even more personal odyssey, one in which psychoanalysis would now play an explicit part.

Post-colonialism, psychoanalysis, and the Popular Front

Wright was not the only African-American to choose resistance over affirmation, to be sure. We have already cited the archetypal progenitor of this choice, Margaret Garner, the slave who killed her infant daughter. According to Joel Williamson, the first black militant in American history, Sam Turner, appeared almost contemporaneously with *The Souls of Black Folk*, wiping out 17 police in Atlanta, Georgia before he was killed. In general, as I have stressed, resistance presupposed affirmation, as suggested by the black young man who confronted his lynchers with the words "Joe Louis." By the time Wright left the Communist Party in the middle of World War II, the psychological, cultural, and political seeds of the Civil Rights movement were apparent in struggles to end discrimination in hiring and to desegregate the military. As the war ended, Westerners learned that the concentration camps had reproduced the logic of slavery, but at a higher technological and lower moral level: aiming to reduce the human being to *zoon*, bare life, just as slavery had, but killing its victims even more quickly than the slave masters had.

As we saw earlier, African-American cultural reconstruction began as part of what I have termed the modern idea of culture: the idea that "man" was a symbol-producing animal, that what constituted the human community were relations of recognition, and that culture was the core of a people's identity and their means of entry into the stream of human progress. Wright, of course, was a critic of this idea, locating violence, trauma, and resistance, rather than cultural achievements per se, at the center of African-American history. In the post-war context, especially after leaving the Communist Party, Wright reformulated his overall idea of culture in a way that put resistance at its center. This effort involved him with many currents of modernist thought. "We know how some of the facts look when seen under the lenses of Marxist concepts," he wrote after leaving the Party, "but the full weight of the Western mind has yet to bear upon this forgotten jungle of black life. What would life on Chicago's South Side look like when seen through the eyes of a Freud, a Joyce, a Proust, a Pavlov?"

The key to understanding Wright's efforts to answer this question lies in his move to Paris in 1947. I said before that the Popular Front defined itself by what it was not, namely fascist. Since fascism itself had an uncanny character, as shown in its familiarity, its recognition as something that had been seen before, and in the horror that accompanied this recognition, the Popular Front can be said to have been trying to negate a negation. In moving to Paris, Wright was escaping the worst depredations of American racism, but he was also separating himself from his *Heim* – the African-American community. In becoming an exile, a cosmopolitan, and a post-Marxist, then, Wright was also occupying the position Freud called *Unheimlich*, uncanny, that is the "not-home" that haunts the home. His move to Paris, in other words, was a deepening, as well as repudiation, of the Popular Front.

From his *Unheimlich* position, Wright created his greatest character, already present in *Native Son* and running through all his subsequent work, the character of the outsider, the individual who is at the border between social inclusion and exclusion, and who is both charismatic and the target of violence. Without abandoning the concrete character of race and racialism, Wright also began to see the Negro as one expression of a more general category, someone who, Wright wrote, lives "in one life, many lifetimes," and who, though "born in the Western world, is not quite of it."[33] For Wright, the outsider spoke to the primal trauma or emptiness at the heart of society, the trauma of slave abductions, lynching, and rapes, that runs through Western history. At the same time, Wright also began to argue that *outsiders* – Negroes, criminals, underground men, metaphysical rebels, *homo sacers* – precisely because they were least integrated into society, offered a new, if still inarticulate, entry point for unraveling and reconstituting the modern idea of culture. In Freud's essay on the uncanny, Freud suggests that the original source of the childhood terrors from which the experience of the uncanny proceeds is the primal father. In Freud's later work, guilt is the outcome of the murder of the primal father, and therefore every individual's fate. This vision of primal guilt, restated in existential terms, was a core strand in Wright's image of the outsider.

"The Man Who Lived Underground," published in 1942, then rewritten and published again two years later, was the first clear indication. The story of a black man falsely accused of murder, living in an underground sewer, observing reality "from below," the work is a brilliant exploration of the phenomenology of blackness, indeed, of any situation of exclusion, insofar as exclusion reflects the origins of society in persecution. Although the police have framed the main character, he accuses himself, describing himself as "always trying to remember a gigantic shock that had left a haunting impression upon one's body which one could not forget or shake off, but which had been forgotten by the conscious mind." When one of the police finally expresses a willingness to accompany him into the sewer, his eyes shine and his heart swells with gratitude: the policeman "believed him ... He had triumphed at last! At last he would be free of his burden." Almost ecstatic, he leads the police down, underground, into the sewer where the "currents" are strong. There the police shoot him: "we are not going to follow that crazy nigger down that sewer are we?" Earlier they had asked him about a statement he had been forced to sign confessing to murder, but now they say "He *is* the statement." The story portrays the tragic double bind of the scapegoat, whose exclusion establishes an illegitimate reign of law and whose fate is foretold in his unconscious.

Wright also used the image of the outsider to revisit Du Bois's notion of doubleness. Du Bois, by describing the Negro as a "seventh son" who lies behind a mystical "veil," had evoked a salvationist, redemptionist role for the African-American community, which was still a quasi-religious role, reflecting the mind-cure provenance of Du Bois's image. Wright, by contrast, substituted psychology for religion. In his 1954 novel, *The Outsider*, the District Attorney – disabled, and thus an outsider himself – explained,

> Negroes, as they enter our culture, are going to inherit the problems we have, but with a difference. They are outsiders and they are going to *know* that they have these problems. They are going to be self-conscious; they are going to be gifted with a double vision, for, being Negroes, they are going to be both *inside* and *outside* of our culture at the same time.... They will become psychological men, like the Jews.

This is a very different notion than the "dialectical overcoming" of *The Souls of Black Folk*. No longer preoccupied with that work's cancellation of racial hierarchy, preservation of difference, and elevation of hybridity, it points toward a revolutionary trans-valuation of the modern idea of culture.

Because Africans had been used like animals, even slaughtered (but not sacrificed), because they had been denied the free status that characterized the human, because they were treated as "not really human," they were outsiders. Exclusionary violence, then, was the premise of culture, not a remediable defect. With this insight, Wright contributed to the inversion of the modern idea of culture. The African slave, he argued, was not someone who had been excluded from a progressive telos of recognition. Rather, the exclusion of the slave was the precondition for the idea of culture, intersubjectivity and mutual recognition. Culture, then, was not a conscious process of recognition but rather a defensive screen, somewhat akin to a dream in that it worked over violence and trauma to make itself acceptable to the conscious ego. Here one begins to glimpse, for the first time, a *truly* modern – actually late modern – idea of culture, one that begins to accept responsibility for the traumatic acts that constitute modernity, and that have been compulsively repeated, as during World War II. Perforce, such an idea rested on a psychoanalytic worldview. Culture, in Wright's new framework, was not the aspiration through which human beings distinguished themselves from animals, but rather, the site of "a fatal division of being, a war of impulses," a "screen" that men and women use to divide "that part of themselves that they are afraid of [from] that part … they want to preserve."

Wright's interest in psychoanalysis flourished in the context of this rethinking, but always as part of his larger project of using literature to explore the possibilities of modern freedom. Like Dostoevsky, the major literary influence on his work, Wright pursued the theme of the outsider by studying the so-called criminal mind, always the main subject of his fiction, and the best laboratory in which to study violence. Spending a great deal of time at courts, jails, and halfway houses, he developed extensive relations with two psychoanalytically oriented psychiatrists who specialized in crime: Frederic Wertham, director of the Mental Hygiene Clinic at Bellevue and author of *Dark Legend*, a psychoanalytic study that explained a case of matricide in terms of unconscious guilt, and Benjamin Karpman, a psychiatrist at St. Elizabeth Hospital in Washington, D.C. who taught at all-black Howard University and, like Wertham, had a special interest in race. Wertham may have treated Wright, as well as assisting him in an ill-fated intervention in the New Jersey prison system, while Wright appealed to various foundations and mayor's committees for funds to publish Karpman's case studies of black criminals, studies that Karpman hoped would prove the "scientific parallel to *Native Son*." But whereas Wertham and Karpman were interested in scientifically understanding the relations of race and crime, Wright used his novels to explore the moral dilemmas faced by individuals under oppressive conditions.

Wright's immersion in the psychoanalytic milieu of the 1940s, as well as the special slant he brought to that milieu, is apparent in his frequent use of Freudian ideas and terminology. In 1943, for example, while traveling through the South on a Jim Crow train with the black sociologist Horace Cayton, co-author with St-Clair Drake of *Black Metropolis*, Wright insisted that the two men eat in the dining car. The white steward sat them last, in the least desirable table, pulling a curtain around them so no one could see them. After they ate, Wright queried Cayton about their black waiter: "Did you notice that waiter when he talked to the steward? ... Poor black devil, his voice went up two octaves and his testicles must have jumped two inches into his stomach ... He does that to emasculate himself, to make himself more feminine, less masculine, more acceptable to a white man."[34] Similarly, in his last novel, *The Long Dream*, Wright wrote of the Southern black father, Tyree: "No white man would ever need to threaten Tyree with castration; Tyree was already castrated."

The androcentrism of such passages echoed that of American psychoanalysis of the 1940s. But, for Wright, working close to unconscious material, androcentrism was sometimes leavened by bisexuality. In his remarkable short story, "Man of All Work," the hero, Carl, had been a cook in the army but he is now busy taking care of a newborn and a sick wife; the only paid work available is as a domestic. To feed his family and save his home, Carl dresses as a woman and gets a maid's job. The story plays on the intertwined irrationalities of gender and race. Not only is Carl invisible as a black man, his cross-dressing performance is violated when the head of the house tries to rape him. The story's high point comes when the wife of his employer forces Carl to give her a bath, thus requiring that he look upon a naked white woman. At its end, Carl's whole family – a newborn, a 6-year old, and a sick wife – are crying and Carl joins in: "Aw, Christ, if you all cry like that, you make me cry ... Oooouuwa." Wright's recognition of the significance of aggression for black masculinity was sometimes balanced by a feeling for male tenderness, vulnerability, and familial involvement.

In 1945 Wright forged further connections between race and psychoanalysis by joining Wertham in opening a low-cost therapy center, the Lafargue Clinic, in Harlem. The clinic, an important experiment in culturally oriented "mass therapy," charged 25 cents an hour and did not pay the psychiatrists. Ralph Ellison called it the "most successful attempt in the nation to provide psychotherapy for the underprivileged" and "one of the few institutions dedicated to recognizing the total implication of Negro life in the United States." General Omar Bradley, director of the Veterans administration, recommended it to all veterans, regardless of race. Wright wrote numerous articles popularizing the clinic, including "Psychiatry in Harlem," in which he spoke of the "artificially made psychological problems" of blacks.[35] His interest in psychoanalysis as a practice also reflected the influence of his friend, Horace Cayton, who had undergone, and written about, a long personal analysis. Beginning with the idea that race was a "convenient catchall," a rationalization for personal inadequacy, a "means of preventing deeper probing," Cayton concluded that race "ran to the core of [his] person-ality," and "formed the central focus for [his] insecurity." "I must have drunk it in with my mother's milk," he remarked.[36]

The decolonization that followed World War II precipitated a shift in Wright's thought from a perspective centered on the United States to a fully global conception of culture, one that took in the pathologies of the west as well as the traumas of the colonized world. As I have suggested, the starting point was Wright's emigration to Europe. In 1945 Ralph Ellison wrote him:

I've been reading some fascinating stuff out of France concern-ing plays written and produced there during the Occupation.... Kierkegaard has been utilized and given a social direction by a group who have organized what is called 'Existential Theater': and, from what I read, their existential probing has produced a powerful art. France is in ferment. Their discussion of the artist's responsibility surpasses anything I've ever seen... They view the role of the indi-vidual in relation to society so sharply that the leftwing boys, with the possible exception of Malraux, seem to have looked at it through the reverse end of a telescope. I am sure that over there the war has made the writer more self-confident and aware of the dignity of his craft. Sartre, one of the younger writers, would have no difficulty understanding your position [i.e., antagonism] in regard to the Left.[37]

Wright lived in France from 1947 until his death in 1960. Influenced by the anti-colonial struggles of the French left, he applied his intuitions concerning trauma and the outsider to decolonization. Invited to the 1955 Bandung Conference of Non-Aligned states, he asked the psycho-analytically oriented social psychologist Otto Klineberg, famous for his studies of "psychological disabilities" of American Negroes, for a series of questions to put to Africans and Asians. In 1957 Wright wrote:

The elite of Asia and Africa are truly men without language .... It is psychological language that I speak of. For these men there is a 'hole' in history, a storm in their hearts that they cannot describe, a stretch of centuries whose content has been interpreted only by white West-erners: the seizure of his country, its subjugation, the introduction of military rule, another language, another religion – all of these events existed without his interpretation of them.... the elite has no vocabu-lary of history. What has happened to him is something about which he has yet to speak.[38]

Beginning in Mississippi with his childhood discovery of the individual's capacity to negate all messages coming from society, Wright had
been led to a view of history that had negation at its center. Realizing
that modern history had a "hole" in it, "a storm in the heart," he was
led to explore the significance of inarticulateness.

By that time – the cusp of the New Left – psychoanalysis had left a
deep mark on African-American cultural thought. W.E.B. Du Bois recognized this when his 1896 doctoral dissertation, *The Suppression of the
African Slave Trade* was reprinted. In an "Apologia" he wrote:

> The work of Freud and his companions and their epoch-making con-
> tribution to science was not generally known when I was writing this
> book, and consequently I did not realize the psychological reasons
> behind the trends of human action which the African slave trade
> involved. Trained in the New England ethic of life as a series of con-
> scious moral judgments, I was continually thrown back on what men
> 'ought' to have done.[39]

Since that time a whole series of black intellectuals have concurred.

Overall, to conclude, the transformation of African-American cultural thought parallels the history of the blues. The emergence of the
blues is almost a miracle: no one knows who invented it or where
the songs were first sung and played. All we know is that they spread
throughout black America in the late-nineteenth century, representing
a dawning awareness of the tragic history of slavery, and its persistent
reverberations. Still, black America did not remain content with the
blues. A deeper descent into the black past entailed the exploration
of dissonance, diminished thirds, fifths, and sevenths, minor scales,
tragic modes, semiquavers, deceptive cadences, caesuras, dominants,
negras, contras, dirges, minstrelsy, flats, taps, scats, and silences, as
well as exchanges with other musical traditions: Italian, Portuguese,
Asian, Indian, Middle Eastern, Jewish. Over time, the blues transformed
nineteenth-century American music – Protestant hymns, Scotch-Irish
ballads, and German marching bands – into the incomparable riches
of bebop, ragtime, soul, rap, jazz, gospel, rhythm, and blues, and rock
and roll, as well as American popular music and musical comedy. An
artifact of the "American Century," the blues became the "most important single influence on the development of Western popular music,"
giving the twentieth-century West that quality of "soul" that Du Bois
first recognized.[40] Ultimately, however, African-American music moved
beyond the blues.

Those who have suffered the kind of traumatic depredations that African-Americans suffered want and need to tell their story, perhaps more than they want and need anything else. Often, however, those who have suffered passively and repeatedly over long periods of time, come to believe that they are themselves at least partly at fault, and so they lose the capacity to speak. Psychoanalysis, even though, and perhaps even especially because, it was far removed from the direct experience of most African-Americans, exemplified the effort to give voice to traumatic suffering, and this effort inspired Wright's life work. As we excavate the modern idea of culture, then, the transnational impact of psychoanalysis in moving modern cultural thought beyond the blues should be recognized.

Notes

1. Joel Williamson, *The Crucible of Race* (New York: Oxford University Press, 1984).
2. Paul Gilroy, *Black Atlantic* (Cambridge, MA: Harvard University Press, 1983).
3. Ron Eyerman, *Cultural Trauma* (New York: Cambridge University Press, 2001) p. 2.
4. Tom Nairn, *The Break-up of Britain: Crisis and Neo-Nationalism* (London: New Left Books, 1977, 1981) pp. 348–349.
5. Joel Williamson, *Crucible*, p. 403.
6. Paul Anderson, *Deep River: Music and Memory in Harlem Renaissance Thought* (Durham, NC: Duke University Press, 2001) p. 6.
7. Frances Lee Utley in Robert E. Hemenway. *Zora Neale Hurston: A Literary Biography* (Urbana, IL: University of Illinois Press, 1977) p. 8.
8. Hemenway, *Zora Neale Hurston: A Literary Biography* (Urbana, IL: University of Illinois Press, 1977) p. 114.
9. Henry Louis Gates Jr., *The Signifying Monkey: A Theory of Afro-American Literary Criticism* (New York, NY: Oxford University Press, 1988) pp. 170–216.
10. Jean Toomer, "Negro Psychology in *The Emperor Jones*," In Robert B. Jones (ed.), *Jean Toomer: Selected Essays and Literary Criticism* (Knoxville, TN: University of Tennessee Press, 1996) p. 6.
11. Werner Sollors, "Jean Toomer's Cane: Modernism and Race in Interwar America," In Geneviève Fabre and Michael Feith (eds), *Jean Toomer and the Harlem Renaissance* (New Brunswick, NJ: Rutgers University Press, 2001) p. 20.
12. Zora Neale Hurston, *Moses, Man of the Mountain* (New York: Lippincott, 1939).
13. Ralph Ellison, "Richard Wright's Blues," In John F. Callahan (ed.), *Collected Essays of Ralph Ellison* (New York: Modern Library, 1995) pp. 128–133, discussing Wright's *Black Boy*.
14. W.E.B. Du Bois, "My Evolving Program," quoted in Claudia Tate, *Psychoanalysis and Black Novels: Desire and the Protocols of Race* (New York: Oxford University Press, 1998) p. 51.

15. Edward Margolies, *The Art of Richard Wright* (Carbondale, IL: Southern Illinois University Press, 1969) p. 55.
16. I am following here Abdul R. Janmohamed in Henry Louis Gates and Anthony Appiah, *Richard Wright: Critical Perspectives* (New York: Amistad, 1993).
17. Richard Wright, *Black Boy* (New York: Harper Collins, 1945) p. 284.
18. Richard Wright, *The Long Dream* (Chatham, NJ: Chatham, 1958) p. 165.
19. Wright, *Black Boy*, pp. 271–272; Margaret Walker, *Richard Wright: Demonic Genius* (New York: Warner, 1988) p. 41.
20. St. Clair Drake and Horace Cayton, *Black Metropolis* (New York: Harcourt Brace, 1945) Introduction.
21. Quoted in Donald Gibson, *Five Black Writers* (New York: New York University Press, 1970), pp. 24–25.
22. Wright, *Black Boy*, p. 372.
23. Arthur Koestler *et al.*, *The God that Failed* (New York: Bantam Books, 1949, 1958) p. 118.
24. Mark Naison, *The Communists in Harlem* (New York: Grove Press, 1984) p. 211.
25. Wright, *Black Boy*, p. 135; Koestler, *God that Failed*, p. 131.
26. Hazel Rowley, *Richard Wright: The Life and Times* (New York: Henry Holt, 2001) pp. 136–137.
27. Wright is quoted in Gates and Appiah.
28. At his death, Wright's personal library included more than a dozen books by Freud, as well as works by Wilhelm Reich, Wilhelm Stekel, and Theodor Reik, mostly in editions from the late thirties and early forties.
29. E. Fuller Torrey, *Freudian Fraud* (New York: Harper Collins, 1992) pp. 35–37.
30. Quoted, Rowley, *Richard Wright*, p. 33.
31. Ellison, "Richard Wright's Blues."
32. Irving Howe, "Black Boys and Native Sons," In Irving Howe (ed.), *A World More Attractive* (Freeport, NewYork, Horizon Press, 1970, original publication 1963), pp. 98–122.
33. Wright, 1953 quoted in Cedric J. Robinson, "The Emergent Marxism of Richard Wright's Ideology," *Race and Class* (Vol. XIX, no. 3), p. 227.
34. Horace R.Cayton, "The Search for Richard Wright," (1969), Box 4, Cayton Collection, Chicago Public Library.
35. "Psychiatry in Harlem," *Time*, December 1, 1947 and "Clinic for Sick Minds," *Life*, February 28, 1948.
36. Horace R. Cayton, *Long Old Road* (New York, NY: Trident Press, 1965) p. 260. See also Horace R. Cayton, "Personal Experience of Race Relations," unpublished 1967 paper, Box 3, Cayton Collection, Chicago Public Library.
37. Ralph Ellison to Richard Wright, July 22, 1945 quoted in Michel Fabre, "Richard Wright, *French Existentialism and the Outsider*," quoted in Yoshinobu Hakutani, *Critical Essays on Richard Wright* (Boston, MA: GK Hall, 1982) p. 184.
38. Richard Wright, *White Man, Listen* (New York: Doubleday, 1957) p. 36.
39. W.E.B. Du Bois, *The Suppression of the African Slave Trade* (New York, NY: Russell and Russell, 1965) pp. 327–329.
40. Stanley Sadie, ed., *The New Grove Dictionary of Music and Musicians* 2nd edn. (New York: Oxford University Press, 2000) Vol. 3, p. 730.

Section 2
Psychoanalysis and Transnational Politics

3
Primitivity, Animism and Psychoanalysis: European Visions of the Native 'Soul' in the Dutch East Indies, 1900–1949

Frances Gouda

As a small European democracy in Northern Europe, the Netherlands had achieved political and economic mastery of a large and lucrative colonial empire in Southeast Asia. During the decades before and after 1900, the Dutch empire in the Indonesian archipelago was "rounded off from Sabang to Merauke," that is, from the Sabang harbor on the northwestern tip of Sumatra to the town of Merauke on the eastern-most border of Dutch-controlled territory on the island of (Papua) New Guinea. The Netherlands' self-described role as *gidsland* (guiding force or guiding light) in colonial affairs – "a Cunning David amidst the Goliaths of Empire"[1] – renders twentieth-century Dutch East Indies scholarship concerning ethnic cultural customs and conventions (*adat*), Islam and animist religions and the vagaries of the "native soul" particularly interesting. Because of a Dutch desire to project itself as international and a progressive pioneer in the proper management of European colonial power in Asia, its possession of the enormous Indonesian archipelago placated the "oversensitivity of a small nation with a heroic past" and substantiated its claim to be a mouse that could still roar.[2]

As in other European countries and the United States, the psychoanalytic ideas of Sigmund Freud began to percolate in the Netherlands during the first decades of the twentieth century. As elsewhere, professional psychiatrists and physicians were internally divided in their responses to Freud's new insights, which resulted in a lively scholarly debate. Due to the interconnectedness of the Netherlands' national identity with its role as an imperial superpower in Asia, however, the new psychoanalytical emphasis on the role of the universality of the unconscious in the human experience also entered ethnographic discourses in the world of Dutch culture overseas. Aside from the political entanglements between mother country and its Southeast Asian colony, the Dutch East Indies in the twentieth century continued to be a lucrative source of income as well, generating between 15 and 20 percent of the Netherlands' gross national product until 1942. Since the turn of the century, most Dutch civil servants received a thorough academic training in Indology at the University of Leiden, where they were instructed in indigenous languages, ethnography, *adat* law and Muslim and animist religions. As a result, the transnational exchange of ideas between the Dutch Metropole and its colonial empire incorporated the flow back and forth *not* only of people, money and goods but also of new ideas about the relationship between body and soul and the (un)conscious mind in understanding local cultures as well as the aetiology of psychiatric afflictions.

The reception of psychoanalysis in the Netherlands

The influential Dutch writer Frederik van Eeden, who also happened to be a physician, was inspired by Eduard von Hartmann's metaphysical notions about spiritualism and psychic monism in the second half of the nineteenth century. As a result, he became fascinated with emerging ideas about the relationship between the materiality of the body and the more unfathomable aspects of the human mind. His intellectual fascination prompted him to pay a visit to Jean-Martin Charcot's theatrical demonstrations of hysterical patients in Paris in November 1885. Van Eeden's personal odyssey from positivist medical science to a more profound understanding of the connections between physiology, human conscience and neurosis prompted him to open a clinic for psychotherapy (later named *Instituut Liébeault, inrichting voor psychotherapie ter behandeling van zenuw – en zielsziekten* or Liébeault Institute for the treatment of disorders of the nervous system and the soul).[3] He founded the institute in Amsterdam in 1887 together with his colleague Dr. Albert-Willem van Renterghem, where they began to treat patients with hypnosis.[4]

In one of the first of a long series of writings in 1888, entitled *De psychische geneeswijze* (psychic medical treatment), Van Eeden formulated his evolving vision of human unconscious: "Our intellect, our potential to think rationally, functions as the senate of the republic or as the general staff of the army. It is here that contemplation and reasoning take place; [it is in the conscious mind] that messages are received, decisions are made and commands are issued." He continued to observe, however, that deliberate thought processes comprise only a small element in the functioning of the human conscience. "At any given moment, we are rarely aware of our thoughts and actions; most of what happens inside us transpires in the great sphere of either the semi-conscious or complete unconsciousness... [and] the unconscious element inside us actually constitutes an entirely different human being, an entirely different I (*een geheel ander ik*)."[5]

Because of his burgeoning spiritualism and mysticism, which entailed a growing renunciation of the realities of the flesh and sexual desires, Frederik van Eeden eventually emerged as an ambivalent critic of psychoanalytic practices. He echoed the negative opinions of Cornelis Winkler, a prominent professor of psychiatry at the universities of Utrecht and Amsterdam. Winkler dismissed psychoanalysis as nothing but a "house of cards... an elegantly written fantasy" that comprised no more than a "mediocre novel or a scholarly chimera;" instead, he had argued, it was the scientific analysis of the nervous system and human reflexes that constitute the primary goal of modern psychiatry.[6] From a different perspective Van Eeden began to fault psychoanalysis for linking mankind's search for "higher" spiritual transcendence within a process of curbing and sublimating "lower" physical urges. In 1910 he wrote in his diary that "this man [Freud] claims to a dangerous degree to be the hero who has liberated us from materialism. But he turns out to be much worse than the materialists."[7] He called Freud a "cynical, crude spirit" (cynische ploertige geest) and judged psychoanalysis a contamination, or a "suffocating vapor" (verstikkende walm) due to its emphasis on sexuality.[8] According to Van Eeden, sexuality belonged primarily to the inferior realm of animals, which prompted psychoanalytic theory to deny the superior and transcendental qualities of human beings: "Freudianism itself is an illness, a psychosis of psychiatrists... who burrow with their sick minds into the innocent souls of unfortunate patients and help to spread disease" instead of curing it.[9] In 1913 he attended the International Psychoanalytic Congress in Munich as an observer, not as a formal participant, along with other literary figures such as the poet Rainer Maria Rilke. During the heated discussions at the Munich psychoanalytic gathering, venting the smoldering conflict between Freud and Jung, Van Eeden spoke on behalf of spiritual transcendence and the "uplifting nobility of humankind." In his diary he later described his performance in Munich as a "nightmare" because in his defense of spiritualism, he proceeded to dismiss psychoanalysis as a series of "grotesque obscenities that pretend to be scientific." He also noted afterwards that he found the manner in which the "attentive men and women at the conference absorbed all this filth with sanguine faces... [to be] genuinely demonic" (waarlijk demonisch).[10]

Soon thereafter, however, Van Eeden changed his opinion yet again after a visit to Vienna and a lunch meeting at Freud's house, followed by an exchange of letters about the nature of "lucid dreaming" and the Great War.[11] From 1914, he ceased to write negatively about Freud but maintained his skepticism with regard to psychoanalysis as both a therapeutic practice and a vision of the human experience. Because of Frederik van Eeden's contested but towering presence in Dutch politics, culture and intellectual life during the early twentieth century, however, his sometimes antagonistic engagement with psychoanalytic ideas may have reinforced their currency. While most professional psychiatrists in this era tended to cling to a positivist or materialist diagnosis of mental illness as grounded in physiological causes, there were others, such as Gerbrandus Jelgersma, Leendert Bouman, Arie van der Chijs and the brothers Johan and August Stärcke, who initiated a progressive and open-minded psychiatric discourse in the Netherlands in which Freud's and Jung's ideas were seriously considered, debated and gradually incorporated. As Van der Chijs wrote in 1914 in his *Inleiding tot de grondbegrippen en techniek der psychoanalyse* (Introduction to the Basic Concepts and Techniques of Psychoanalysis), "the *libido* is a life force containing all the human yearnings for satisfaction, happiness and love." This libido, however, should not only be controlled or sublimated but above all rendered useful in order to apply it toward productive ends. People should be liberated from cultural restrictions that functioned as "a corset that distorts the human spirit" by encouraging a sense of shame through the treatment of sexuality as "stepchild" of society. Psychoanalysis, he noted, is not a multi-headed monster that sows misery or degeneracy among those who fall within its clutches; rather than leading to promiscuity, psychoanalysis "brings only truth."[12]

For the first generation of devotees of Freud and Jung in the Netherlands, the new science promised to be a harbinger of cultural transformation and an innovative medical discourse that might be capable of liberating the prudish social sensibilities embedded in Dutch society. In the eyes of a Dutch group of readers of Freud and Jung, psychoanalysis could ideally serve as a tool in unmasking Victorian hypocrisies and, in due course, produce honest new forms of modern subjectivity and openness toward sexual morality in the Netherlands. They founded the *Nederlandse vereniging voor psychoanalyse* (Dutch Association for Psychoanalysis) in 1917, which joined the International

Psychoanalytic Association soon after its establishment in 1920. Also in 1917 a sophisticated scholarly debate about psychoanalysis appeared on the pages of the foremost medical journal *Nederlandsch tijdschrift voor geneeskunde*, that is, the Dutch Journal of Medicine. Many articles in this medical quarterly in 1917 recorded the lively interest among a range of physicians, psychiatrists and neurologists, who discussed the relative merits of Freud versus Jung and offered insights into how psychoanalysis could enhance not only the treatment of mental illness but also foster the psychological well-being of modern society at large. Freud's induction as an official honorary member into the *Nederlandse vereniging voor psychoanalyse* in 1921 registered the receptivity to psychoanalytical approaches among a progressive sector of the Dutch psychiatric establishment.

Psychoanalytic influences in diagnoses of the "native soul" in the Dutch East Indies

In the Dutch East Indies, Freudian notions concerning the universal unconscious indirectly influenced physicians and ethnographers who were preoccupied with understanding the psychology of the indigenous population of Java and other regions of the Indonesian archipelago. During a gathering in early 1908 of the Indies Association (*Indisch Genootschap*) focusing on the topic of the position of native physicians (*dokter Djawa*) trained at the medical school in Batavia, the Dutch psychiatrist, anthropologist and anatomist Johan Herman Frederik Kohlbrugge addressed his fellow medical doctors, colonial administrators and prominent representatives of the private sector. At the outset, Kohlbrugge stated his conviction that anatomically, the essential characteristics of the Javanese mind and psyche were identical to those of human beings in the modern Western world.

What remained to be resolved, however, was his observation that the Javanese possessed a character and personality that differed from people in the West while also maintaining a different relationship to the natural environment. These distinctions were combined with a darker skin color. Although he seldom mentioned the issue of skin color in his voluminous writings on the Javanese anatomy and 'soul,' he noted on occasion that the golden glow – or rather, a lack thereof – of Javanese complexions constituted an outward sign of inner psychological turmoil. These distinctions, however, did not render the Javanese inferior (*minderwaardig*) but as "equal in value" albeit different (*gelijkwaardig, anders*). Due to an abiding belief of even the most sophisticated

Javanese people in animism, trying to teach them the facts and concepts of modern Western science would be fruitless, for the time being, unless their faith in animism would be replaced with monotheistic religion, whether it be Christianity or a genuine form of Islam. As long as the Javanese, even as European-trained doctors, persisted in their mental reliance on magical incantations, ritual objects and religious teachers and healers (*ilmu, gnelmu, talisman, dukun, guru*) in order to assuage the unpredictability and dangers of powerful spirits of nature, the Javanese psyche would remain infantile and conform to the earliest childhood of human history.[13] In this context Kohlbrugge added elsewhere that unruly Javanese children were never disciplined or punished because of the parents' fear a child might lose its delicate "zielestof" (semangat, the soul's spiritual matter or its spiritual potency).[14]

Because Javanese society placed a premium on the maintenance of outward composure and calmness as a visible sign of refined (*halus*) status, psychologically the Javanese were an inherently tense people out of fear of failing to meet these injunctions. As he wrote in his *Een en ander over de Psychologie van den Javaan* (This and That about the Psychology of the Javanese) in 1907, a lack of individuality caused the average person in Java "to live in a perpetual state of nervous terror concerning the vengefulness of both natural spirits and fellow citizens." Their reliance on fate produced among the Javanese a kind of sullen determinism, a credo of "non-being" ("niet zijn" in the original), the destruction of his own self or ego (*eigen ik*) and of "thinking without real thoughts." Referring to the diagnosis of Dr. Van Brero, a consulting physician in the psychiatric hospital in Buitenzorg (Bogor) who had hosted the research visit of the famous European psychiatrist Emil Kraepelin in February and March of 1904, the most often-occurring mental disease among the Javanese was acute amentia. In Kohlbrugge's view, this illness was generated by "auto-suggestion provoked by an overheated body due to malaria," which caused a loosening of psychological associations. Amentia was expressed in "frantic mobility...tearing up clothes, making everything dirty, dancing, jumping, shouting, roaming through the countryside, and occasionally in murderous outbursts and incoherent language" (i.e., *amok*).[15] "As is the case among Englishmen," Kohlbrugge opined, Javanese preoccupations with the projection of serenity to the outside world "is a deceptive and acquired trait" that was not automatically transmitted from one generation to the next, as Lamarckian evolutionary theory might propose.[16]

As the son of a prominent Calvinist theologian, Kohlbrugge was well-versed in Biblical scholarship. Using the "comparative method" to highlight the debilitating impact of animism on the human psyche and group morality, he skipped over Greek and Roman civilizations in classical antiquity to return to the ancient Israelites for the purpose of illustrating his point of view. In the book of Judges in the Old Testament, he noted, some of the great leaders of the Israelites tried with divinely ordained spiritual might to bind their followers to monotheism, that is, the faith in Jahweh (spelled as Jahve). But the people regularly strayed from this belief by "prostituting" themselves to animism again (the Dutch verb he used was *hoereert*, which was also the term used in the official Bible translation issued by the Estates General of the Dutch Republic in 1637).[17] During those regressive intervals, the prehistoric Israelites worshipped beasts, mountains, trees or the anthropomorphized forces of nature with names such as Baäl, Dagon and Astaroth. And every time the Israelites forsook monotheistic religion, he argued, they lost their collective strength and were conquered by surrounding animist tribes, all of whom either hated or envied the monotheism of the ancient Hebrews. Again defeated and subjugated by animist neighbors, the Israelites recovered the memory of their monotheistic religion as it had sustained their previous communal resilience. Kohlbrugge conceded that this result should not come as a surprise, because ethnographers everywhere had shown that animism fostered "particularism," while monotheism reinforced social solidarity and communal feelings. As intellectual ammunition he quoted appreciatively the published research of the sophisticated Dutch-reformed missionaries Nicolaas Adriani and Albert Kruyt in Northern Celebes (Sulawesi), whose ethnographic findings confirmed the lack of individuation among the animist Toradjas (*Toraja*). Citing once more the example of the ancient Israelites, Kohlbrugge noted the powerful influence of women who, in his view, functioned as what Freud would label "atavistic vestiges." He therefore repeated his warning that in the Dutch East Indies, "our experiment [to teach Western science to Javanese doctors] will not be successful as long as the Javanese mother has not been liberated from animism."[18]

The writings of the prolific J.H.F. Kohlbrugge provide a useful starting point for thinking about the ways in which Dutch East Indies psychiatrists, anthropologists and ethnographers in the period between 1900 and 1949 were conversant with the ideas of biological and cultural evolution, on the one hand, and Sigmund Freud's psychoanalytic re-articulations of evolutionary theories, on the other hand. Did Dutch analysts of the Javanese 'soul' – to repeat the title of one of Kohlbrugge's major books, *Blikken in het zieleleven van den Javaan en zijner overheerschers*

(Glimpses into the Soul of the Javanese and the Soul of his Rulers) – consciously partake of international discussions concerning evolutionary theory formulated by Jean-Baptiste de Lamarck, Charles Darwin and Ernst Haeckel, debates that were reconfigured and given renewed currency in Freud's critical psychoanalytic project in the twentieth century? Did they express their views on the so-called infantile psyche of the Javanese, steeped in the superstitions of animism and beliefs in magic and the ominous spirits of nature, by drawing upon the foundational literature of the modern academic disciplines of anthropology and sociology by Edward Burnett Tylor, James Frazer, Herbert Spencer and William Robertson Smith, Emile Durkheim, Gustave le Bon, or Wilfred Trotter?

This is a salient question from the perspective of transnational history through the vectors of European imperialism in Asia. Dutch colonial governance in Southeast Asia tried to distinguish itself from other European colonizing nations. By means of its emphasis on finely crafted anthropological insights, a profound knowledge of Islam, and historical and jurisprudential research on the different forms of local *adat* (ethnic customs and traditions), the Dutch colonial establishment in Europe and Southeast Asia cultivated its uniqueness in relations between West and East. As Professor Johan Christiaan van Eerde, the director of the anthropology division of the Colonial Institute in Amsterdam, formulated it in 1914, "a little country such as Holland provides better guarantees than larger nations to implement the appropriate policies. Bigger European countries display a tendency to use brute force in colonial administration – a blunt violence that is grounded in their self-assurance as societies that can wield superior political and military might." Dutch colonial practice revealed, instead, that "the greater the weakness" of a particular country in the international arena, the more it would try to nurture the cultural evolution of native people and "to acknowledge and accommodate the cultural predilections of members of indigenous societies." Little Holland was not able to rely on "*domme kracht* or *plomp geweld*" (a reckless sense of strength) to impose its will and determination on native people. Accordingly, Van Eerde emphasized that the Netherlands, instead, had always recognized and nurtured the distinctive, if fragile, nature of a wide range of indigenous cultures, which made it an exemplary colonizing power.[19]

It was, therefore, likely that in the course of interpreting the findings of their ongoing ethnographic research on *adat*, culture, psychology, animism and Islam, scholars and anthropologically inclined missionaries in the Dutch East Indies were well-aware of the international literature on cultural and biological evolution, whether written in English, German or French, that percolated in the Europe metropole and radiated to its imperial possessions during the early-twentieth century. Indeed, judging from Kohlbrugge's wide-ranging writings, one may surmise that he was conversant with the implications of Edward Burnett Tylor's evolutionary approach to understanding, categorizing and taxonomizing primitive cultures around the globe. His use of the ahistorical "comparative method" in deploying the ancient Israelites in order to shed light on the animist psyche and sexual unruliness of the Javanese in the early-twentieth century, whether sophisticated doctors or simple farmers, is a striking example and makes a plausible case. A rigorous diachronic or topographical unfolding of history was obviously not Kohlbrugge's concern, nor was it an issue for the evolutionary anthropologists who also fueled Freud's imagination. Kohlbrugge probably took his cue from Tylor's foundational text of the modern academic discipline of anthropology, *Primitive Culture. Researches into the Development of Mythology, Philosophy, Religion, Language, Art and Custom* (1871), where he wrote that "little respect need to be had in comparisons [of primitive cultures] for date in history or place in map; the ancient Swiss lake-dweller may be set beside de medieval Aztec, and the Ojibwa of North America beside the Zulu of South Africa."[20]

Also Kohlbrugge's argument that animism was a "cultural survival," which was recapitulated in successive generations, hints at a familiarity with Freud's psychoanalytic adaptations of the biological evolutionary ideas of Lamarck, Darwin and Haeckel. The passages cited above also suggest he may have been aware of William Robertson Smith's *Religion of the Semites* (1894), which figured in Freud's *Totem and Taboo* (1913) as well. Robertson Smith had argued that behind the positive religions of Judaism, Christianity and Islam – grounded in the teachings of great religious innovators, who spoke as instruments of divine revelation – lies the "old unconscious religious tradition." These unconscious animist traditions expressed connections between the gods and their worshippers "in the language of human relationships . . . taken with strict literality. If a god was spoken of as a father and his worshippers as his offspring, the meaning was that the worshippers were literally of his stock, that he and they made up one natural family with reciprocal family duties to one another."[21] Kohlbrugge developed a discourse concerning Javanese selfhood or the soul of *natuurmenschen* (tribal people) in general that engaged Freud's emphasis on the homology between primitivity and the infantile stages of human development in the West or, as in the title of the final chapter of *Totem and Taboo*, "the return of totemism in children."[22]

In 1913, at a time when the academic discipline of anthropology in England and the United States had begun to wean itself, both literally and figuratively, from biological and cultural evolutionary models under the influence of the theory of cultural relativism of Franz Boas and the functionalism initiated by Bronislaw Malinowski, Freud published *Totem and Taboo: Resemblances between the Psychic Lives of Savages and Neurotics*. In this book he articulated his meta-historical visions of the relationship between the primitivity of the many uncivilized "savage and half-savage" tribes that populated the world's "dark continents," on the one hand, and the origins of neurosis experienced by civilized adults in the West, on the other hand. In *Totem and Taboo*, Freud set forth his perspective on evolutionary theory and the nature of history that had inflected the scientific *Zeitgeist* of the *fin de siècle*. He wrote:

> What is older in time is more primitive in form...Primitive man is known to us by the stages of development through which he has passed...and through the remnants of his ways of thinking that survive in our own manners and customs. Moreover, in a certain sense he is still our contemporary. There are [contemporary] people whom we still consider more closely related to primitive man than to ourselves, in whom we therefore recognize the direct descendants and representatives of earlier man. [The mental life of savages] assumes a peculiar interest for us for we can recognize in their psychic life a well-preserved picture of an earlier stage of our own development. If this supposition is correct, a comparison of the psychology of primitive peoples as taught by social anthropology with the psychology of the neurotics as it has become known through psychoanalysis, will reveal numerous points of correspondence and throw light upon familiar facts of both sciences.[23]

In *Totem and Taboo*, Freud articulated in psychoanalytic detail an opinion concerning evolutionary theory he had already succinctly formulated in earlier work. In the *Interpretation of Dreams*, which appeared in 1900, he had observed that "behind an individual's childhood we are promised a picture of a phylogenetic childhood – a picture of the evolution of the human race, of which the individual's development is in fact an abbreviated recapitulation influenced by the chance circumstances of [his or her] life." He continued to argue that "Dreams and neuroses seem to have preserved more mental antiquities that we could have imagined ... so that psychoanalysis may claim a prominent place among the sciences concerned with the reconstruction of the earliest and most obscure periods of the beginnings of the human race." In *Three Essays on the Theory of Sexuality*, published in 1905, he noted that "the phylogenetic process can be seen at work in the ontogenetic process." This ultimately entailed a rehearsal of "earlier experiences of the species to which the more recent individual experience, as the sum of accidental factors, is super-added." In his 1911 *Psychoanalytic Notes on an Autobiographical Account of a Case of Paranoia (The Schreber Case)*, Freud maintained that "In dreams and neuroses ... we come upon the savage *(den Wilden)* ... upon primitive man as he stands revealed to us in the light of the researches of archaeology and of ethnology."

Following *Totem and Taboo*, the analogies between the narcissistic and Oedipal phases of individuals in childhood and the evolutionary stage of animism in primitive tribes recurred frequently in Freud's writings. In a paper in 1914 focusing "On Narcissism," for instance, he equated the widespread beliefs among dark-skinned tribal people that ritual and magic might placate supernatural, "uncanny" (unheimisch) forces in the natural world with a child's overestimation of its own emotional desires and mental acts in efforts to alleviate the uncertainties of the domestic environment. According to Freud, both womanhood and childhood constituted "atavistic vestiges." He argued that a child often harbors feelings of omnipotence, wishful thinking or a blind faith in "the thaumathurgic [miraculous] force of words." This narcissism, sometimes combined with a primordial form of megalomania, emerged from efforts to placate or influence the unpredictability of a child's surroundings, just as savages used spiritual incantations, rituals and magical objects in their attempts to mitigate and mollify the mysterious spirits of the natural landscape.[24]

In *Group Psychology and the Analysis of the Ego* in 1921, Freud acknowledged – but contested and reframed – the sociological insights of Gustave le Bon and Wilfred Trotter, who had analyzed the conduct of human beings, mobilized in mass movements, by linking the collective behavior of the group to a lack of individuality that was reminiscent not only of primitive peoples but also of the herd instincts of animals such as sheep.[25] He implied that every modern mass movement – reinforced by the nefarious institutional powers of the State and the Church in modern Europe – re-enacted the "dynamics of a primitive community whose prototype was the primal horde in thrall *(Hörigkeit)* with an authoritative patriarchal leader."[26] Finally, in *Civilization and Its Discontents* which appeared in 1930, he again elaborated on the same historical analogies:

> When a child reacts to his first great instinctual frustrations with excessively strong aggressiveness and with a correspondingly severe superego, he is following a phylogenetic model and is going beyond the response that would be currently justified; for the father of pre-historic times was undoubtedly terrible, and an extreme amount of aggressiveness may be attributed to him.[27]

Concluding with his posthumously published book on *Moses and Monotheism* in 1939, Freud's work is replete with references to Darwinian themes as well as Haeckel's controversial ideas about ontogeny reca-pitulating phylogeny and Lamarck's contested notions concerning the inheritance of acquired characteristics. In all of his work, Freud showed he was immersed in the fields of biology and anthropology. Darwin's evolutionary theory, grounded in a biological process of sexual selection and adjustment to environmental challenges that were instrumental in the competitive survival of the "most favoured races" – as he noted in the subtitle of the *Origins of Species* – became one of the pillars of the project of psychoanalysis, anchoring Freud as a "biologist of the mind," as Frank Sulloway has called him.[28] Freud was also enam-ored with the evolutionary insights of Lamarck, arguing in favor of the transmission of acquired characteristics from parents to children. His *Philosophie zoologique*, originally published as early as 1817, experienced an enthusiastic revival as the implications of Darwin's evolutionary ideas began to sink in and mutate in late-nineteenth century European scholarly circles. Haeckel's recapitulation theory became a touchstone of the psychoanalytic project as well, although Freud probably disagreed with Haeckel's statement that primitive people such as the Australian Aborigines were closer to apes and dogs and their lives, therefore, should be assigned a lower value. On the whole, Freud was fascinated with questions about the relationship between what Karl Marx had called the historical tension between "species *being* and species *life*" – in other words, the connection between the social particulars of an individual life and the evolving structural constraints of capitalism – reverberating in late-nineteenth century social-democratic and communist move-ments in Germany, Austria and Russia as well as in France, England and the Netherlands. In *Totem and Taboo*, Freud incorporated these various strands of scholarly and political ideas that circulated and intermin-gled in the early-twentieth century, and the 1913 book constituted "the culmination of all that goes before and was the starting point for all that comes after."[29]

In the Dutch East Indies, a range of Kohlbrugge's fellow psychiatrists and anthropologists echoed views that make it credible to argue that they linked their findings to wider international debates on cultural evolution and psychology. As a Dutch psychiatrist in Java, Dr. P.H.M. Travaglino, alleged in 1920 the simple and naïve populations of Java, Sumatra or Bali belonged to the same species as white-skinned Europeans, but they embodied an immature and still malleable stage of human development. Adult Javanese men and women exhibited the psychological foibles of children, because "the natives (*inlanders*) are still in an earlier stage of their evolutionary development."[30] It is unlikely that this argument emerged in a vacuum. He was fascinated with the psychological implications of the homology between "primitivism and infantility," if only because it might shed indirect light on the early formation of neurotic European personality structures.[31]

Travaglino argued that so far, the environmental circumstances that enveloped the average Javanese adult had prevented the progressive, natural advancement of higher forms of intelligence and more exalted cultural impulses. Their instincts, he wrote, "are younger in an evolutionary sense," because, over time, the gradual adaptation of the human species to their physical environment bestowed upon mankind a greater capacity for rational thought, more elevated passions and refined, subtle sensibilities. But among the Javanese, instead, all sorts of pre-pubescent "primary instincts" predominated – "the yearning for pleasure and the excessive influence of sexuality" – which Travaglino interpreted as evidence of the puerile nature of the Javanese psyche. The mentality of the Javanese, he asserted, despite the cultural values placed on calmness and placidity, was characterized by "an infinite capacity for unrestrained emotionality, a strong fantasy life, and occasionally an obsessive concentration of their attention" on a particular object. Travaglino also claimed that adult Javanese men and women suffered from "a limited *vigiliteit*" (watchfulness or vigilance), which seemed to be a direct contradiction of their stubborn tenacity.[32]

Thus the Javanese, in terms of the evolutionary process, represented the developmental level of children (*kinderlijk niveau*), since the structural conditions of their lives had prevented, to a great extent, the natural maturation of more sophisticated impulses, which were of a more recent vintage in an evolutionary sense. The Javanese, he concluded, "have progressed at a slow pace in the evolutionary process (*langzaam voortschrijden in het evolutionaire proces*)."[33] Javanese people, in other words, possessed the same mental "essences" as Europeans but had become stuck, for the time being, at the level of childhood, which explained the simultaneously different mentalities of East and West.

Infantile associative thinking, a thoughtful Dutch ethnographer named F.D.E. Van Ossenbruggen maintained, was prevalent almost everywhere in the Indonesian archipelago. Any person with a "naive mind," he argued in 1925, was easily satisfied with "striking resemblances, which are substitutes for positive, convincing evidence." They treat "visual analogies" between a wide variety of objects, or subjective resemblances between two divergent experiences, as tangible properties or "identities" and adjust their behavior accordingly.[34] In Java and among many ethnic groups in the Indies, Van Ossenbruggen cited as one among many other examples, that people envision the spread of smallpox as an illness caused by "an ugly man, a negro or an old woman, who propagates the disease by sprinkling fruit pits or little peas around themselves." This "crude, materialistic, associative thinking" suggested that individuals could save themselves from being infected with smallpox simply by avoiding close physical contact with homely men, people with black skins, elderly women, peas and fruit pits. He wondered, though, whether the "associative thinking" of primitive people differed substantially from the analogic thought processes of rational, Western men. "A child's blushing cheeks remind us of red apples," he noted impishly, whereas a European peasant, whose village neighbor had emigrated to America and had become fabulously rich, often expected that analogous wealth automatically awaited him purely because he hailed from the same hamlet in the Old World and moved to the same, distant continent.[35]

In general, Dutch psychiatrists suspected that "mental illness... hardly existed among primitive peoples" in Java. Their views were bolstered by Emil Kraepelin's judgments after his visit to the psychiatric institution in Buitenzorg in 1904.[36] As a diagnostic expert on dementia praecox and manic-depressiveness – in due course renamed schizophrenia and bipolar disorder – he had examined its prevalence among the Javanese, Chinese and European hospital residents. In comparison to European patients, his conclusion was that mental illness not only occurred less often and in a milder form in Javanese patients, also their prospects for recovery were better. Secret voices in the head and visual hallucinations were seldom reported to Kraepelin and other contemporary or subsequent Dutch physicians. Mental health professionals in twentieth-century Java tended to interpret the manic and depressive phases of bipolar disorder as a by-product of the distinctive nature of Javanese social relations and personhood, with its emphasis on communally internalized forms of serenity and decorum. Because of the Javanese appreciation for the outward appearances of soft-spoken gentility, withdrawals into depressive silence or manic outbursts of *amok* were translated into a respectful Dutch discourse on the particular tensions embedded within Javanese culture and personality rather than conveyed in a diagnostic vocabulary of psychiatric afflictions. In summary most Dutch physicians portrayed aberrant behavior and mental instability as an inherent feature of the primitivity, infantility and animism of the Javanese psyche constrained by the status injunctions of Java's culture.

Freudian analysis as intellectual ammunition against Indonesian nationalism

In the interbellum era, when the Indonesian nationalist movement gained political strength, new interpretations of the Javanese psyche were produced. There was no unanimity in Europeans' views; instead, the Dutch colonial community began to broadcast a cacophony of voices that was modulated by factors such as political perspective, social background and level of education. In the course of the 1920s and especially during the 1930s, a palpable paranoia about Indonesian nationalism eclipsed the intellectual anarchy of competing opinions. Whether Dutch colonial inhabitants approached the colonized "other" as an unformed child, an underdeveloped species or a cultural survival of a medieval European "self," the white-skinned community began to refashion the Indonesian elite with whom they maintained direct contact more and more as quasi-intellectuals, who manipulated the peasant masses for their own "childlike" (*kinderlijk, kinderachtig*) and selfish political purposes.[37] In the hearts and minds of a growing proportion of Dutch East Indies' residents, these pseudo-educated upper-class natives began to resemble cocky demagogues – adolescent "roosters who think that by crowing noisily they can accelerate daybreak."[38] Not yet mature enough to realize that a little Western knowledge could hardly transfigure them into commanding officers of their independent ship of state, native politicians' irrational conviction that they were capable enough to replace their Dutch colonial masters confirmed their infantility and privileged one particular strand of evolutionary thinking. It was in this new political climate prior to and immediately after World War II that a Dutch psychiatrist, Dr. Pieter Mattheus van Wulfften Palthe, initiated a new discourse concerning the Javanese psyche, one that increasingly embraced a psychoanalytic point of view.[39]

During the first half of the twentieth century, Freud-inspired concepts such as megalomania, masochism or revulsion had already emerged as subversive descriptions of the colonial system. The thoughtful Indonesian nationalist Sutan Sjahrir, for instance, wrote in the 1930s that colonial mastery had always fostered "sadism and megalomania" among Dutch East Indies civil servants and settlers – the very same Dutchmen who tended to be reasonable and tolerant human beings if they lived in either Amsterdam, Rotterdam or The Hague. The unnatural qualities of the colonial condition, Sutan Sjahrir noted, caused unpretentious Dutchmen to indulge in delusions of grandeur and omnipotence once they reinvented themselves as white-skinned "lords of the manor" in the Indonesian archipelago. As the acerbic Indonesian nationalist Tan Malaka had commented earlier, "every ne'er-do-well Dutch wastrel was a potential *grand seigneur* in Deli" (i.e., in the rubber and tobacco plantation economy on the east coast of Sumatra).[40] In the process, these unassuming Dutchmen, once transplanted in Javanese or Sumatran cultural soil, almost automatically proceeded to deny the subjectivity of the indigenous populations by saddling "Indonesians' gentle souls with inferiority complexes."[41]

In the immediate post-war period, Dr. Van Wulfften Palthe used Freudian ideas to psychoanalyze the Indonesian side of the colonial predicament. He noted that in Javanese language and culture, a distinction existed between the concepts of culture (*boedaja*) and civilization (*adab*); the latter he considered to be the wide range of powerful civilizing restraints and inhibitions imposed on Javanese individuals in order to maintain their equanimity and refinement. He defined culture as the sum total of a society's "possessions in mental and spiritual values: religion, art, science and technology, law and political science." These "products of the human mind" are inculcated and obtained through individual effort and they nurture and complement each other; together they constitute the cultural characteristics of a given community. He argued, however, that culture can't exist without civilization, and that the "passive and polite mask" of the average native person did not imply "apathy or lethargy" but functioned, instead, as the concealment of a "fierce sensibility" which occasionally ran *amok* and found expression in sudden outbursts of "wild aggression."[42]

When the civilizing norms that regulate and command a community's collective behavior are "introjected, then the Super-Ego is formed which, now acting from within, becomes responsible for good manners and moral conduct." But for the Western-educated members of the native elites, the situation was more complicated and made their character less stable. In early childhood, during the "building up of their Super-Ego, the commands, inhibitions, examples and ideals of their own Oriental environment were introjected," prompting children in the Indonesian archipelago to identify with their own father's image. However, starting at the age of six they began to be exposed to Western influences. Rather than reinforcing the "original Super-Ego," however, European education "created a new one, which took its place beside the first." Van Wulfften Palthe called this a dualistic or "a sort of double Super-Ego" that existed independently but contiguously because "it" was – or "they" were – not integrated and frequently provided contradictory injunctions and instructions. Owing to this "double-mindedness," the impulses unleashed by the Ego and the Id in Western-educated intellectuals in Indonesia imposed on them "greater and still more varied self-denials, which again led to more restrictions and frustrations."[43]

Van Wulfften Palthe noted that in the patrimonial colonial society of the Dutch East Indies, Indonesian intellectuals had viewed the average Dutchman as a "father-imago," which implied the polar emotions of love and hate and caused a conflict of loyalties with their Oriental Superego. But in 1942, the "militaristic typhoon of a small yellow race" descended from the north and swiftly toppled the Dutch East Indies government. The independent Indonesian nationalist and communist, Tan Malaka, commented in his diary on the easy defeat of the Netherlands Indies military forces in 1942, though, he made the Dutch colonial father figures seem like kernels of "sand blown from the rocks" by a gentle breeze rather than a mighty storm.[44] Japan's effortless victory prompted Van Wulfften Palthe to argue that it destabilized Indonesians' "child–father sentiments," which they quickly transferred from their Dutch fathers to the new Asian rulers from Japan.[45]

During the Japanese occupation, however, Van Wulfften Palthe noted that Indonesians eventually redirected their feelings of rancor, too, perhaps because the Japanese were more hardhearted as "fathers" than the Dutch had been. The Japanese had not only inherited but also improved upon the "instruments of oppression" left behind by the Dutch, Tan Malaka recorded in his journal, which induced fear and loathing.[46] Although it may be an apocryphal story, a Dutch report in 1944, citing an American diplomatic source, claimed that the inhabitants of Java paid obeisance to Japanese military officers during the morning hours by transforming the formal Japanese greeting *Ohayo gozaimas(u)* into *Hajo gasak mas*, meaning "come on, let's kick him."[47]

In the wake of the Western allies' defeat of imperial Japan in August, 1945, Van Wulfften Palthe argued that the Indonesian population could no longer attach their "polar sentiments" to a particular object. All father figures had disappeared and Indonesian men's emotional polarities suddenly existed in a vacuum, "deranging the structure of their super ego" by emphasizing the negative, spiteful extremes of their emotional bipolarity.[48] Initially, catchphrases such as *merdeka* (freedom, independence) with its accompanying pseudo-Nazi salute, as the Dutch psychiatrist called it, or red-and-white flags, functioned as fetishistic alternatives to the positive self-image Indonesians had previously derived from the paternal approval of either Dutch or Japanese officials.[49] Indonesians' revolutionary slogans – *rakyat* (the people) or *manusia* (humanity), *perjuangan* (struggle), *kedaulatan* (sovereignty), *semangat* (dynamic spirit or spiritual potency) and of course *merdeka* (independence, freedom) and *revolusi* (revolution) – began to function as magical credos, which expressed a newly forged intra-ethnic solidarity and produced an "ideological intoxication."[50]

Such "mantras" or "mystical chants" also bolstered the united purpose of Indonesians, even across the great social divide that existed between exalted *priyayi* members of prominent families whose power and social status could be identified by the particular ending of their last names and simple peasants in the countryside.[51] Given Van Wulfften Palthe's views concerning the inherent emotionality of the Indonesian people, however, and assuming that hatred tends to function as an obstinate element in the human heart, he theorized that in the psychology of Indonesians, "a gradual fallback set in toward a primitive stage of development."[52] Or, in the language of a Dutch pulp novelist in 1947: *merdeka* began to operate as "an incomprehensible, magical word without content," a term that disclosed all "the primitive bloodthirsty instincts of hatred."[53]

The Dutch psychiatrist asserted that these unique forms of regression, even among Western-educated Indonesians who had achieved mental adulthood, produced "infantile reactions" in the affective sphere. This make-believe world of children gave "free play" to the most elementary impulses of aggression and cruelty, while in the intellectual sphere, "archaistic-intuitive reasoning" began to prevail and wish-fulfillment superseded logic and rationality. In a comparable fashion, a former Dutch prime minister dispatched to Java to negotiate with representatives of the Indonesian Republic in 1946 to 1947, Willem Schermerhorn, noted in his diary that many Indonesians suddenly seemed to live in "an eternal dream world, in which no one dared to call a spade a spade."[54]

According to Van Wulfften Palthe, a patricidal urge came in the wake of "secondary narcissism" and "infantile reasoning", and Indonesians' ambivalent emotions "fell apart into their polar factors." On the one hand, the positive desire to please the father figure was replaced with a mother fixation – a set of feelings Indonesians began to express in a tempestuous new love for *Ibu pertiwi*, "mother earth," who symbolically embodied the Republic.[55] He also alleged that many Indonesians found a new love object in their own personalities as a form of "secondary narcissism," while all destructive emotions of anger and hatred were now projected onto Western politicians, Dutch soldiers, and "NICA-bandits" (NICA was the Netherlands-Indies Civil Administration) bent on vanquishing the Republic.[56]

Conclusion

Van Wullften Palthe's psychoanalytical account of Indonesia's struggle for independence constitutes evidence of the domestication of Freudian discourses in the post-World War II era. Freud's literary and analytic skills in writing about the interstices between biology and evolutionary theory, between individual history and group psychology and between anthropology and archeology, produced an innovative language and intellectual mindset that settled in the popular imagination of the twentieth century. Eli Zaretsky, in his recent *Secrets of the Soul: A Social and Cultural History of Psychoanalysis* (2004) has described it as a vocabulary that was particularly suited to the Fordist and post-Fordist culture of the twentieth century with its emphasis on individual personhood and modern subjectivity, or its consumerism and reliance on the appeal of advertising. Even more so, Freud's psychoanalytic idiom was appropriated by the 1960s, a decade that has become known as the beginning of the "culture of narcissism" in the modern Western world.[57]

In one of Claude Lévi-Strauss's later publications entitled *The View from Afar* (1983), in which he provided a sweeping overview of the field of anthropology in the twentieth century, he introduced a chapter on "Cosmopolitanism and Schizophrenia" with a psychoanalytically inspired narrative, even though during his active anthropological career he had routinely and skeptically interrogated the value of psychoanalysis. In 1949, for instance, he had labeled Freud's emphasis on the resemblances and trans-historical overlap between primitive mentalities and the psychological dilemmas of childhood in the modern Western world as a "tempting but spurious correspondence" that emerged from what he referred to as "the archaic illusion" of psychoanalysis. He had argued, instead, that in any given culture, those who don't conform to prevailing social norms and forms tend to be dismissed as childish and primitive, which does not necessarily constitute "a return to an archaic 'stage' in the intellectual development of either the individual or the human species."[58] But in *The View from Afar*, written later in 1983, he paid homage to Freud's insights by introducing the syndrome of schizophrenia, for example, as "an oscillation between two extreme feelings: the insignificance of one's ego in relation to the world, and the overweening importance of oneself in relation to society."[59]

With regards to the ethnographic and psychiatric analyses of the Javanese soul in the twentieth century Dutch East Indies, it is therefore plausible to conclude that in the period between the writings of J.H.F. Kohlbrugge and P.M. Van Wulfften Palthe, a gradual immersion in psychoanalytic discourses took place, even if it was used to as a weapon to depict the behavior of Indonesian nationalists as pathological and thereby dismiss the legitimacy of the independent Indonesian Republic. One could even argue that in the colonial world of the Dutch East Indies, the evolutionary writings of Edward Burnett Tylor, Sir James Frazer and fellow-travelers in their anthropological cohort received a second purchase and a new hearing through a re-articulation of their ideas in psychoanalysis. In a certain way, this conclusion should not come as a surprise. Whether in a Dutch, English, French or other colonizing context, white-skinned residents were always searching for a rhetoric of legitimacy capable of normalizing or "naturalizing" the subjection and exploitation of indigenous peoples. Franz Boas's theories on cultural relativism or Bronislaw Malinowski's functionalist approach to anthropology were less useful than the evolutionary taxonomies of Tylor, Frazer, Lubbock, Morgan and Maine in furnishing Europeans with a discourse of justification. Psychoanalysis gave evolutionary anthropological ideas a renewed currency during the first half of the twentieth century. It is likely that from Freud's point of view, however, this was one of the unintended consequences of his intellectual opus.

Notes

1. The title of chapter 2 of Frances Gouda, *Dutch Culture Overseas: Colonial Practice in the Netherlands Indies, 1900–1942* (Amsterdam: Amsterdam University Press, 1995), pp. 39–74.
2. P.M.B. Blaas, "De prikkelbaarheid van een kleine natie met een groot verleden: Fruin en Bloks nationale geschiedschrijving," *Theoretische geschiedenis*, Vol. 9, No. 2 (1982), pp. 271–304; translated quote in Gouda, *Dutch Culture Overseas*, p. 23.

3. Harry Stroeken, *Freud in Nederland. Een eeuw psychoanalyse* (Amsterdam: Boom, 1997), pp. 17–18.
4. Ilse N. Bulhof, *Freud en Nederland* (Baarn: Amboboeken, 1983), pp. 67–78. See also Christine Brinkgreve, *Psychoanalyse in Nederland: een vestigingsstrijd* (Amsterdam: Arbeiderspers, 1984).
5. Frederik van Eeden, "De psychische geneeswijze," *De Nieuwe Gids*, Vol. 3 (1888), pp. 383–433, quoted by Bulhof, *Freud en Nederland*, pp. 70–71.
6. Stroeken, *Freud in Nederland*, p. 25.
7. H.W. van Tricht, ed., *Dagboek, 1878–1923*, IV volumes (Culemborg: Tjeenk Willink-Noorduyn, 1971), vol. II, p. 1111, cited by Bulhof, *Freud en Nederland*, p. 94.
8. Bob Rooksby and Sybe Terwee, in "Freud, Van Eeden and Lucid Dreaming," translate this statement as "a cynical coarse soul." See www.spiritwatch.ca (accessed August 8, 2007).
9. van Eeden's, *Dagboek*, vol II, p. 1120, cited by Bulhof, *Freud en Nederland*, pp. 94–95.
10. van Eeden's, *Dagboek*, vol III, p. 1333, cited by Bulhof, *Freud en Nederland*, pp. 94–95.
11. Rooksby and Terwee, in "Freud, Van Eeden and Lucid Dreaming," note that Ernest Jones's biography of Freud reports that Van Eeden was "an acquaintance of his old hypnosis days;" Freud's letters to Van Eeden were written on March 14 and December 28, 1914. See www.spiritwatch.ca (accessed August 8, 2007).
12. Quoted by Bulhof, *Freud en Nederland*, pp. 228–229.
13. J.H.F. Kohlbrugge, vergadering van januari 28, 1908, "Iets over de inlandsche geneeskundigen," in *Verslag van het Indisch Genootschap* (The Hague: Indisch Genootschap, 1908), pp. 123–124.
14. J.H.F. Kohlbrugge, *Een en ander over de psychologie van den Javaan* (Leiden: E.J. Brill, 1907), pp. 50–52.
15. Ibid., p. 42.
16. Ibid., footnote 1, p. 44. Throughout his life, Kohlbrugge presented himself as an opponent of evolutionary notions, proposing that the descent of humankind occurred without some kind of transcendental divine force. As he asked in his conclusion to his *Critiek der descendentietheorie* (Utrecht: N.V.A. Oosthoek's Uitg. Mij, 1936. Critique of the Theory of Descent) in 1936: "where [what] is the leading force that produced diversity, logical structure, utility? ... One can either posit the existence of an inherent force dedicated to variation or a transcendent protoplasm, or one can contest concepts such as growing power, adjustment or entelechy. In the latter case, one may either use the word X or the word God" p. 126.
17. With gratitude to Dr. Peter van Rooden for alerting me to the original seventeenth century Bible translation.
18. Kohlbrugge, "Iets over de inlandsche geneeskundigen," pp. 126, 129.
19. J.C. van Eerde, "Omgang met inlanders," in *Koloniale Volkenkunde*, mededelingen no.1 (1914; Amsterdam: J.H. de Bussy, 1928), p. 54.
20. Edward Burnett Tylor, *Primitive Culture. Researches into the Development of Mythology, Philosophy, Religion, Language, Art and Custom* (1871; New York: Brentano, 1924), p. 6.

21. William Robertson Smith, *Burnett Lectures on the Religion of the Semites* (1888–1889; London: Adam and Charles Black, 1894), pp. 1, 29–30.
22. See, among others, J.H.F. Kohlbrugge, *Primitieve denkwijze I & primitieve denkwijze II: uitstralende krachten in de volkenkunde* (Amsterdam: Koninklijk Nederlands Aardrijkskundig Genootschap, 1920), vol. XXXVII, pp. 729–755; *'S Menschen religie; inleiding tot the vergelijkende volkenkunde* (Groningen: J.W. Wolters, 1932); *Critiek der descendentietheorie* (Utrecht: NV A.Oosthoek's Uitg. Mij, 1936).
23. Sigmund Freud, *Totem and Taboo: Resemblances between the Psychic Lives of Savages and Neurotics* (1913; New York: Barnes & Noble, new ed. 2005). Trans. A.A. Brill; Introduction by Aaron H. Esman, p. 1.
24. The quotations of Freud's views on primitivity are based on discussions in the following sources: Eli Zaretsky, *Secrets of the Soul: A Social and Cultural History of Psychoanalysis* (New York: Alfred A. Knopf, 2004); Ranjana Khanna, *Dark Continents: Psychoanalysis and Colonialism* (Durham: Duke University Press, 2003); Celia Brickman, *Aboriginal Populations of the Mind: Race and Primitivity in Psychoanalysis* (New York: Columbia University Press, 2003); Edward W. Said, *Freud and the Non-European* (London/New York: Verso, 2003); Joseph H. Smith & Afaf M. Mahfouz, eds., *Psychoanalysis, Feminism, and the Future of Gender* (Baltimore/London: The Johns Hopkins University Press, 1994); Peter Gay, *Freud: A Life for Our Time* (New York/London: W.W. Norton, 1988); Edwin R. Wallace IV, *Freud and Anthropology: A History and Reappraisal* (New York: International Universities Press Inc., 1983); Géza Róheim, ed., *Psychoanalysis and the Social Sciences*, 3 Volumes (New York: International Universities Press, 1947, 1949, 1951); George B. Wilbur & Warner Muesterberger, eds., *Psychoanalysis and Culture: Essays in Honor of Géza Róheim* (New York: International Universities Press Inc., 1951).
25. Gustave le Bon, *The Crowd* (London: T. Fisher Unwin, 1910; originally published in French in 1896); Wilfred Trotter, "The Herd Instinct and its Bearing on the Psychology of Civilized Man," part 1 and 2, *Sociological Review* (July 1908 and January 1909); *The Herd Instinct in War and Peace* (London, 1915; New York: MacMillan, 1919).
26. Brickman, *Aboriginal Populations*, p. 94; Wallace, *Freud and Anthropology*, pp. 250–251.
27. Quoted by Brickman, *Aboriginal Populations*, p. 77.
28. Frank Sulloway, *Freud: Biologist of the Mind. Beyond the Psychoanalytic Legend* (New York: Basic Books, 1979). See also Lucille B. Ritvo, *Darwin's Influence on Freud: A Tale of Two Sciences* (New Haven: Yale University Press, 1990).
29. Wallace, *Freud and Anthropology*, p. 1.
30. P.H.M. Travaglino, "Het karakter van den inlander," *Tijdschrift van de Politiek Economische Bond*, Vol. 1 (1920–1921), pp. 342–343. See also, "De psychose van den inlander in verband met zijn karakter," *Geneeskundig Tijdschrift voor Nederlandsch-Indië* (Batavia: Javasche Boekhandel and Drukkerij, 1920), LX, No. 2, pp. 99–111.
31. See the discussion in Ashis Nandy, *The Intimate Enemy. Loss and Recovery of Self under Colonialism* (Delhi: Oxford University Press, 1983), p. 13, and Octave Mannoni, "Psychoanalysis and the Decolonization of Mankind," in J. Miller, ed., *Freud* (London: Weidenfeld & Nicholson, 1972), pp. 86–95.
32. Travaglino, "Het karakter van den inlander," pp. 342–343.

33. Ibid., p. 343, also quoted and discussed in great detail in Paul van Schilfgaarde, "De psyche van de Javaan," *Djawa*, Vol. 5, No. 1 (1925), pp. 109–111.
34. F.D.E. van Ossenbruggen, "Het magisch denken van den inlander," In George Nypels, ed., *De Indische Gids* (also the new series of *Tijdschrift voor Nederlandsch-Indië*) (1926), vol. 48, No. I–VI, pp. 290–299. Lecture delivered to students of the *Handelshoogeschool* in Rotterdam, December 1, 1925.
35. Van Ossenbruggen, "Het magisch denken van den inlander," pp. 297–298.
36. Hans Pols, "The Development of Psychiatry in Indonesia: From Colonial to Modern Times," *International Review of Psychiatry*, Vol. 18, No. 4 (August 2006), pp.363–369.
37. Ph.H. Coolhaas, "Ontstaan en groei," in *Wij gedenken. Gedenkboek van de Vereniging van ambtenaren bij het Binnenlands Bestuur in Nederlands-Indië* (Utrecht: Oosthoek, 1956), pp. 62, 70–71. The Dutch words are *kinderlijk and kinderachtig* (translated as childlike).
38. Hendrik Colijn made the actual statement; see the discussion in Bernard Dahm, *Soekarno en de strijd om Indonesië's onafhankelijkheid* (Meppel: Boom, 1964), p. 341.
39. For a detailed analysis of Van Wulfften Palthe's writings, see the conference paper of Hans Pols, "The Hordes and the Disappeared Totem: A Psychoanalytic Commentary on the Indonesian Struggle for Independence," University of Wisconsin-Madison, 2004, and Frances Gouda, "Languages of Gender and Neurosis in the Indonesian Struggle for Independence, 1945–1949," in *Indonesia* (Cornell University Southeast Asia Publications), Vol. 64 (April 1997), pp. 45–76.
40. Tan Malaka, Tan Malaka, In Helen Jarvis, ed., *From Jail to Jail*, 3 Vols (Athens: Ohio University Press, 1991), Vol 1, p. 45. Helen Jarvis translates this comment as follows: "Every lazy good-for-nothing *schlemiel* who came to Deli from the Netherlands had hopes of becoming a Tuan Kecil, a prospective Deli capitalist."
41. Soetan Sjahrir, *Indonesische overpeinzingen* (Amsterdam: Bezige Bij, 1945), p. 165.
42. P.M. van Wulfften Palthe, *Psychological Aspects of the Indonesian Problem* (Leiden: E.J. Brill, 1949), pp. 39–40. For a Dutch-language publication on the same issues, see *Over het bendewezen in Java* (Amsterdam: F. van Rossen Uitgevers, 1949).
43. Van Wulfften Palthe, *Psychological Aspects*, pp. 41.
44. Tan Malaka, *From Jail to Jail*, Vol. 3, p. 72.
45. Van Wulfften Palthe, *Psychological Aspects*, pp. 9–13
46. Tan Malaka, *From Jail to Jail*, Vol. 3, pp. 72–73.
47. Charles Olke van der Plas, "Situation in the Netherlands Indies," report forwarded to the U.S. Secretary of State by Walter A. Foote from Canberra, Australia on February 26, 1944, Dispatch No. 58, in RG 165, War Department, General and Special Staffs, Military Intelligence Division, Regional File Netherlands East Indies, Box 2631, NA II. Foote added a flourish to his translation of *Hajo gasak mas:* "come on, let's cut off his neck."
48. Van Wulfften Palthe, *Psychological Aspects*, p. 42. For the psychoanalytic background of Van Wulfften Palthe's analysis, see Sigmund Freud, "Splitting of the Ego in the Defensive Process," as discussed by Philip Rieff, ed., *Sexuality*

and the Psychology of Love (New York, 1963), pp. 209–213. Freud had written in 1925 that in women such a derangement of the superego had different implications, because a woman's "superego never becomes so inexorable, so impersonal, so independent of its emotional origins" as a man's, quoted by Gay, *Freud*, p. 516.

49. Takao Fusayama, a Japanese officer in Medan in October, 1945, made a similar observation; he noted that a nationalist leader, while addressing an audience, held "his right hand high like Adolf Hitler, the German dictator," In *A Japanese Memoir of Sumatra 1945–1946: Love and Hatred in the Liberation War* (Ithaca: Cornell University Southeast Asia Publications, 1993), p. 18.

50. Benedict R. O'G Anderson, "The Language of Indonesian Politics," in *Language and Power. Exploring Political Cultures in Indonesia* (Ithaca/London, 1990), pp. 123, 139–140.

51. James Siegel used the words "mantra" and "mystical chants" in *Fetish, Recognition, Revolution* (Princeton: Princeton University Press, 1997), p. 212.

52. Van Wulfften Palthe, *Psychological Aspects*, pp. 13, 42.

53. Piet Korthuys, *In de ban van de tropen* (Wageningen: Gebr. Zomer & Keuning, 1947), p. 204.

54. Van Wulfften Palthe, *Psychological Aspects*, p. 4; *Het dagboek van Schermerhorn*, Vol. 2, p. 695. The Dutch saying is "to call a cat a cat."

55. Van Wulfften Palthe, *Psychological Aspects*, pp. 3, 43–45; Sumathi Ramaswamy, "Virgin Mother, Beloved Other: The Erotics of Tamil Nationalism in Colonial and Post-Colonial India," In *Thamyris: Mythmaking from Past to Present* (1997), elaborates on *Ibu Pertiwi's* counterparts in India, embodied in *Bharata Mata* (Mother India) and regional emblems of what she calls "language/nation/mother," such as *Tamilttaay* (Tamil Mother), vol. 4, No. 1, pp. 9–39.

56. Van Wulfften Palthe, *Psychological Aspects*, pp. 3–4, 9, 11–13, 42–43. NICA was the acronym for Netherlands Indies Civil Administration.

57. Zaretsky, *Secrets of the Soul*, especially Part II entitled "Fordism, Freudianism and the Threefold Promise of Modernity."

58. Claude Lévi-Strauss, "The Archaic Illusion," In *The Elementary Structures of Kinship* (1949; Boston: Beacon Press, 1969), pp. 84–94 and 97.

59. Claude Lévi-Strauss, *The View From Afar* (1983; Chicago: University of Chicago Press, 1985), p. 177.

4

Fascism Becomes Desire: On Freud, Mussolini and Transnational Politics

Federico Finchelstein

> The Id and the super-ego have one thing in common: they both represent the influences of the past.
>
> Sigmund Freud, 1938–1939.[1]

In early 1933 Sigmund Freud received an inconvenient visitor. The visitor was Giovachino Forzano, a renowned fascist opera composer and a personal friend of Benito Mussolini. Forzano's daughter was a patient of Edoardo Weiss, the noted Italian psychoanalyst. A Freudian loyalist, Weiss wanted Freud's personal supervision of the case and he went to the Austrian capital taking with him his patient and her fascist father. The three distinctive individuals showed up at Freud's home on Berggasse 19 on April 26, 1933 and the fascist Forzano asked Freud to dedicate one of his books to Mussolini.[2]

Freud found himself in a difficult position, a double bind of sorts. If he dedicated a book to the Duce, he would be defined as a fascist fellow traveler or worse. But if he decided not to do so, he would probably endanger the already difficult standing of Italian psychoanalysts vis-à-vis the fascist regime.[3]

Moreover, Freud was quite aware of the fact that at this time, Mussolini stood as a "protector" against the Nazis at home (in Austria) and abroad.[4]

Almost three months before Forzano's visit, Hitler had become the Chancellor of Germany. Hitler had full-fledged antisemitic goals or as Freud put it, Nazism meant violence with a program "whose only political theme is pogroms."[5]

Freud wrote in his dedication to Mussolini: "To Benito Mussolini, with regards from an old man that recognizes in the ruler the Hero of Culture."[6]

Did Freud really consider Mussolini his hero? Some years before, in 1928, he had expressed in a private letter, his radical dislike for Mussolini.[7] Many historians have famously presented Freud as being inward looking vis-à-vis Austrian and European politics, and as having an escapist attitude toward politics. As Carl Schorske argues in his classic study, psychoanalysis was born as a result of a Freudian displacement, from the reality of politics to the workings of the mind.[8] To be sure, Freud thought that his "science" of psychoanalysis was a life fulfilling creation that transcended nations, political cultures and identity formations. Thus, he often relegated politics to the contingency of (historical) external life. For him, the study of the workings of the mind linked these historical elements with transhistorical ones. This argument is important and deserves consideration. But there is a political dimension of psychoanalysis that many authors focusing on the "liberal" German elements of Freud's personality tend to downplay.[9]

Freud may have escaped from actual politics but politics did not escape him. There is an emancipatory, almost utopian, dimension in Freud's thinking that includes but also goes beyond his own Austrian national context.[10] Freud had a transnational understanding of global political processes that went beyond restrictive notions of his many identities.[11] It is actually the productive combination of all these subject positions (being Austrian and living in a baroque Catholic environment, being European, being Jewish, being a scientist) that disabled Freud's encounter with identity politics and displaced him to the significations of actual politics. Freud enacted, practically and theoretically, the position of the outsider. He dedicated the last years of his life to trace the political dimensions of desire. His last book *Moses and Monotheism* (1939) epitomizes this Freudian search for meaningful answers in the history of leadership, that is to say, the history of politics. Behind the founding traumas of Judaism, that is behind the symptom, Freud was able to read the political. In this sense, the political dimension was not an absence, an abstract source of metahistorical considerations but a palpable historical loss.[12] It was not an excuse for political retreat but a pathway of engagement.

The political dimension, the Freudian recognition of personal and collective powerlessness in front of fascism, was a central, if not the central, frame of reference for Freud' s last years. To be sure, he actually described in 1933 to Ernest Jones, a personal feeling of numbness which was provoked by "the bleak misery of these times which at present stifles all more meaningful activity for me."[13] But, as I will demonstrate, Freud was not exclusively engaged in melancholic detachment. Throughout

the ideological civil war of the interwar years, Freudian psychoanalysis represented an effort in political understanding. Like anti-fascism at
large, psychoanalysis defined its political place as the result of an act
of disempowerment, namely a subject position affected by the losses
provoked by persecution and victimization. Psychoanalysis early status
as a "pariah," its almost constitutive condition of internal exile and
academic displacement, provided a precedent as well as a conceptual
framework to the anti-fascist perspective of exile. Moreover, it was theoretically suited to analyze fascist aggression as constitutive of the fascist
"idea" of the self.[14] Freud was quite aware, as we will see, of fascist
processes of abjectification and he saw them as radical outcomes of
modernity's propensity to open up to the historical, as well as mythical, forces of desire that had preceded it. Being an expert in the analysis
of myths, Freud confronted the founding political myths of fascism
throughout the interwar years and especially in *Moses and Monotheism*.[15]

Freud's own encounter with Mussolini, Hitler and, more generally, fascism was informed by this active political dimension that Freud could
not escape. Psychoanalysis, in its encounter with fascism, became a form
of anti-fascism. It was not the anti-fascism of the "established" intellectuals but the anti-fascism of the outsiders with no place in society,
culture and politics.[16]

Early on, Freud saw the uttermost consequences of radical desire in
politics. He saw authoritarian personalistic politics as a result of the
reification of affect. The fascist leader is a radical narcissist who wishes
to be loved outside the limits of the law. Fascism provides its own self-
centered definition of transcendence as the politicization of desire – the
will of the leader that is the embodiment of the paternal metaphor. This
idea of the will represents what Freud described as "omnipotence of
thoughts," based on an overestimation of the influence the self "can
exert on the outer world by changing it." Freud argued that this mindset was typical "in our children, adult neurotics, as well as in primitive
people."[17] Like them, and unaffected by the reality principle, fascism
refuses the power of discourse, of dialogue and language, and pro-
poses sacrifice and violence as means and ends for achieving political
desire.[18]

For Freud, the fascist leader thinks in terms of circular images, rei-
fies ritual and radicalizes the political value of performance.[19] There is
much truth to these notions, but the fascist persuasion was for Freud
more importantly embedded in history than in aesthetics or performance. Freud confronted fascism with contextualization. Like Hannah
Arendt, the other great interpreter of fascism, Freud opposed mythical

thinking in politics with the capacity to think.[20] He identified this capacity to think with a combination of critical irony, myths of origins, and more generally with an analytical condensation full of transhistorical elements. Critical irony could be defined as a reassertion of the capacity to think when confronted with a circular vision of the world, a full-fledged totalitarian ideology. Condensation provides the possibility of using analogy by focusing on an object of symbolic and real power. But what are the limits of a criticism when the subject of the critique, the one who is being criticized, cannot understand or even be recognized as such? Do implicit, or even cryptic, political statements such as Freud's "dedication" to Mussolini, represent a form of resistance? Freud provided a conceptual metaphorics of fascism through critical irony and analytical condensation rather than through systematic argumentation. In other words, he presented a language for understanding and surviving fascism, a language full of emancipatory potential.[21]

If fascism put forward a notion of politics as the realm of collective psychology and as an attempt to master individual wills, Freud thought that the fascist attempt to provide closure to political utopias could be only understood in terms of a transhistorical longing to return to a primeval state. Both fascism and psychoanalysis put forward transhistorical, and transnational, notions of political desire. But whereas Mussolini conceived desire, particularly his own, to be a political imperative that transcended history and national territories, Freud conceived fascism as the return of a mythical past, particularly that represented by the myth of Prometheus. He actually put forward this interpretation in his dedication to Mussolini but this was not apparent. I will return to Prometheus in the next section of this essay, but first we will deal with the different intellectual forking paths traversed by Freud's conception of fascism. Literally, these were the Freudian inroads into the understanding of transnational fascism.

When confronted with Forzano's impending fascist request to address the Duce, Freud, a master of reading the implicit, preferred to face off the Duce with apparent praise and encrypted radical critique. Connecting Mussolini with historical examples, or even with explicit Greek myths was out of the question. It would have been too obvious.

Freud gave to Forzano, and Mussolini, a book that was full of intertextual implications. It was a copy of a pacifist book published that year, co-authored with Albert Einstein, the book was entitled *Why War?* Freud gave to Mussolini a book that, as he noted, was forbidden in Nazi Germany![22] To Mussolini, the man who famously claimed that war, in its ultimate accomplishment of radical violence, was the essence of fascism, Freud gave a book that presented war as the reification of death, an example of the "blindness to logic."[23]

Freud personally suggested in its first page that Mussolini was a hero of civilization or as he put it a "cultural hero." In this essay, I provide a close reading of this dedication in terms of its encrypted, and often cryptic, connections to broader dimensions of the Freudian corpus. Equally important, I will put this dedication in context.

Like Mussolini and Forzano did, many historians of fascism had misread Freud's "dedication" to the Duce. They failed, so to speak, to see behind the deliberate or the symptom. By reading Freud's dealing with Mussolini in literal terms they have overlooked the essential contribution of psychoanalytic thinking to the understanding of fascism and, last but not least, they have missed Freud's own contribution to anti-fascism as a political ideology.[24] Was psychoanalysis compatible with fascism? Was Freud sympathetic to fascism as other historians would claim? All these questions are related or even framed by another question which is perhaps more significant: What is the ideological connection between fascism and psychoanalysis? In other words: where these two "philosophies," or focal systems of understanding, affected by each other? In this chapter, I will only address one side of this question, namely the Freudian dimension.

As Eli Zaretsky has suggested, survival and not praise was the overriding motive underlying the psychoanalytic reaction vis-à-vis fascism. In this attempt at surviving fascism, psychoanalysis did not refrain from critically analyzing it. Psychoanalytic critical thinking "was integral to the great coalition that defeated fascism."[25] Psychoanalysis is, and certainly was at its prime time, a form of politico-conceptual anti-fascism.

That psychoanalysis was against fascism, the Nazis knew well. One week after Freud had received Forzano, the Nazis were burning Freud's books all over Germany while stating that Nazism was against the "soul-disintegrating exaggeration of the instinctual life" that psychoanalysis represented.[26] Irony once more seemed to be Freud's best answer to universal fascism. When confronted with Nazi book burning, he stated "What progress we are making. In the Middle Ages they would have burnt me: nowadays they are content with burning my books."[27]

The carnivalization of the outsider transformed his books into real subjects of a textual pogrom. This was a fascist instrumental displacement: Freud, the person, became objectified and his books became sacrificial subjects. But for Freud, fascism was not exclusively, or even principally, medieval. Freud presented the historical condition of fascism as that of a reformulation of the past in the present. Thus, fascism was not derivative of the past but a radical interpretation of it. Hitler, of course, referred to the medieval and Christian tradition of antisemitism as a precursor to his own antisemitism.[28] Freud emphasized the Nazi, and fascist links with Christianity as it had existed in Europe. Europeans had been "'badly christened'; under the thin veneer of Christianity they have remained what their ancestors were, barbarically polytheistic."[29] This was for Freud one of the reasons for fascist antisemitism and like fascism at large, it was rooted in contradictions. As Freud ironically wrote to a disciple exiled in Palestine: "Have you read that Jews in Germany are to be forbidden to give their children German names? They can only retaliate by demanding that the Nazis refrain from using the popular names of John, Joseph, and Mary."[30] The idea of ironic retaliation speaks to the profound sense of critical powerlessness. Only rational engagement and ironic condensation provided solace, and more importantly, understanding.

The general anti-fascist idea that fascism represented the past, namely that it was rooted in a barbaric past, explicitly contradicted Mussolini's famous dictum that fascism, like history, did not "travel backwards."[31] But for Freud, unlike many other anti-fascists, the relationship between fascism and the past was not the result of a mimetic identification or mere derivation. To be sure, fascism was "reactionary" and "medieval"[32] but more importantly it equally presented a novel articulation of the past and its myths. Moreover, Freud saw the fascist connection with the past as a combination of experiences and collective mythical frameworks. Most significantly, Freud believed that humanity's earlier moments were represented by myths. Myths represented the structural foundations of society. Myths represented history before it became properly historical. The past began as a transhistorical reality. In Freud's view of the past, the relation between close and remote is blurred.[33] There is no single date, or period, for a myth insofar as the myth "happened" at the founding moments of human development, namely before history. The myth was a response to the founding trauma of human society.[34] But what was the connection between this idea of the past and Freud's understanding of fascism as rooted in the past? Fascism was not only encompassed by its historical past but also by mythical connotations. For Freud, myths are not so much metaphors for explaining fascism but rather constitute its rooted unconscious.

Fascism presented a tension between its radical nationalism and its transnational dimensions (its imperialist pan-national ideology). Fascist imperialism, for example, is central to any understanding of its transnational dimensions.[35] Although his analysis was essentially transnational, Freud did not consider this dimension. He often presented Nazism as "German fascism" or presented the Austrian fascists as being "cousins" of German fascism.[36] Fascism transcended national borders but for Freud fascist transcendence lay elsewhere. Fascism was a global phenomenon but besides its contextual and generic traits, Freud saw fascism as the transhistorical substantiation of a mythical past. Namely, fascism was a repetition, a novel version of the myth of the primordial father.

The father figure represents the pre-rational, and pre-civilizational word of images, the dominion of the visual over the written word. For example, Nazism's declared anti-Jewish nature, and, last but not least, its repetitive burning of texts by Jewish authors, including the text on war that Freud gave to Mussolini, represented for Freud a confirmation of a long cultural battle between image and language. This was a battle that paganism and Judaism came to epitomize. It is not that Judaism was against any image per se. But the attribution of divine power to a given image runs counter to a longstanding Jewish tradition, namely the sublimation of desire through language. It is language, the text, that allows us access to the sacred. The Jewish injunction against the visual representation of desire was a central dimension of psychoanalysis. For Michael Steinberg, psychoanalysis is as critical engagement that confronts

the duality of a regime of ideology and a regime of representation, whose power and authority are to be penetrated through the counteroffensive of analysis. Manifest content thus cedes its power to the latent at the same time, at least in dreams, that vision and images cede their authority to text. What is latent, what is unconscious, carries too much meaning to be permitted to cross the barrier into the conscious or the manifest without disguise and distortion. Its content is historical violence.[37]

Historical violence, and its visual representation, becomes the object of psychoanalytic critique. Whereas Freud saw Greek myths such as that of Prometheus or contemporary historical leaders like Cesar, Napoleon, Hitler and Mussolini as representing a full aesthetic renunciation of ethics, he considered Moses to represent reason, the triumph of ideas and ethics over performance and images. Moses was, for Freud, a source of intellectual and historical resistance; namely Moses represented life against the forces of destruction.[38] During the same time period the composer Arnold Schoenberg, who shared many affinities with Freud, was also able to perceive in musical subjectivity the same phenomenon that the European ideological war between fascism and anti-fascism represented. Schoenberg did this while describing the (political) conflicts between Moses and his brother in his opera Moses und Aron.[39]

Moses's program was not based on instincts and rituals of violence (such as pre-modern pogroms) but in scripture. In short, it was based on an idea and a concept rather than on an image or a feeling. The image claims to embody the actual presence of the Freudian primordial father whereas language is represented by the displacement of the father figure onto the normative aspects of civilization. The brief rapture of violence, the return of the repressed presented in the breaking of the tables of the law – Moses's violent (irrational) reaction against the image of the divinity (represented in the golden calf) shows for Freud the labile nature of rational engagement. Even Moses was tempted to destroy the books of law, and replaced language with violence. Freud, of course, identified Nazi book burning with mythical atavism rather than with fascist "spiritual" imperialism or Nazi "reactionary modernism."[40]

For Freud, the Christian religion and the armies of Cesar and Napoleon represented a return of the primordial father. But unlike the question of religion and more like Cesar and Napoleon, fascism presented the persona of the hero (the leader) as an immediate presence. Freud was interested in historical leaders, what his disciple and biographer Ernest Jones had called "leaders of men." Accordingly, Moses, Napoleon, Cesar, Hannibal and other leaders epitomize the repetitive return of the father.[41]

As it is well known incest lies at the center of the most important mythical engagement of psychoanalysis: the Oedipus complex. The killing of the primordial feather also features and the two are significantly related.[42] Some historians argue that the Oedipus complex works as metaphor for Freud's own personal retreat from the practice of politics. It should be no surprise then that Freud may have also interpreted politics through this lens. The killing of the primordial father represented a source of emasculation, the "precipitate" reflected in the Greek or Incan myths that allowed normativity and unchecked violence to stop. In short, civilization, or the "cultural process," was based on the renunciation of desire. It rested on the rejection, and displacement, of the desire to kill as well as on the rejection to follow a primordial father, a totemic figure that had tyrannical dimensions. A symbol of this renunciation is the democratic sublimation of the killing of the father as a rejection of strong political leadership. Civilization rests in part on this "cultural frustration."[43]

In Freudian terms, a dictatorship is directed against this democratic sublimation and its consequential renunciation of desire. The return of the father, the violent leader, encompassed a return to "primitive" unmediated violence and the renunciation of norms, namely a rejection of civilization. As a leader, Moses may have been a father figure but Freud presented the existence of many Moseses as proof that mediation and symbolism were not lost but innate to Judaism or, at least, to Freud's own reading of it. Here Judaism, of course, represented a symptom of the possibilities for normative progressive civilization.

Freud saw Christianity as more closely related to paganism than to Judaism. The barbaric Middle Ages were linked to the "radical desire" of the past, rather than with the modernity of the present. The baroque and romantic Catholic resistance against psychoanalysis (and political modernity at large) had more to do with the pleasure principle acted out in the stories of Prometheus, Narcissus and Icarus than with Moses's Judaism.[44] The political totem was also an icon of both desire and image and Freud saw Judaism and psychoanalysis as opposing both. In short, psychoanalytic anti-fascism was as much the ideological front against fascist ideology in the present as well as a confrontation with a barbaric past rooted in unreason, in the incapacity to think and the drive to obey the dictum of desire. This barbaric past, like fascism, was for Freud rooted in the pleasure principle.

The weak utopian dimension in Freudian thinking is to be rooted in the future. In the book he gave to Mussolini, Freud argued: "The ideal condition of things would of course be a community of men who had subordinated their instinctual life to the dictatorship of reason ... But in all probability that is a Utopian expectation."[45]

The choice of the word dictatorship may work as the proverbial Freudian slip but as Louis Althusser would have it with respect to Gramsci, Freud speaks, to the future in the present tense.[46] In other words, he had the capacity to move beyond the pleasure principle, to accept cultural frustration and to leave the barbaric past behind. It was history and politics rather than clinical interest that confronted Freud with the experience of fascism. How can the past become the present? In his *Civilization and Its Discontents*, Freud had stated that it was difficult for him to represent or think about the subject position of a victim of radical violence in the past: a slave, the victim of the inquisition or "a Jew awaiting a pogrom."[47] Suddenly he became a subject in the history of persecution and he soon realized that his own fascist present time uncannily repeated the past. The uncanny nature of fascism – its strange familiarity as a repressed content – may have reminded Freud of the unchecked violence that Jews had suffered in the past and the need to resist it.[48] This inability of modern society to accept frustration led to the dialectical return in the present of the heroic primordial father of the past. Fascism then constituted a structural repetition across time and national borders.

Fascism, the pleasure principle and how Mussolini became Prometheus

The personal connections between fascism and psychoanalysis began not only in a simple therapeutic way but as dialogical challenge to the fascist ideological emphasis on monological exchange. Was Mussolini another historical case? What was the relationship between Mussolini, the subject of Freud's dedication, and the myth that formed Freud's analogical frame of reference? Mussolini the patient was "treated" with a highly sophisticated irony that encrypted him as a radical subject for the psychoanalytic couch. But for Freud, Mussolini was not a "normal" neurotic patient but a historical one and a "hero." But what kind of hero?

One year before his visit to Freud, Forzano had co-authored a play in three acts. The subject had been Napoleon and the co-author was Mussolini.[49] Freud probably was informed of Forzano and Mussolini's play on Napoleon and, more generally, about the Duce's tendency to think himself as a heroic figure rooted both in political myths and history. Mussolini often considered himself a new Bonaparte and a new Cesar. As we have seen, Bonaparte had been a figure of central importance to Freud as well. Freud had regarded the Corsican as the leader who, in historical times, had returned politics to the figure of the primordial father.[50] But Freud did not want to tell Mussolini that he was Bonaparte. Besides, Mussolini would have taken this characterization as a clean compliment. The generic idea of a cultural hero was more cryptic and messier but it equally conveyed a reading of the fascist unconscious that was certainly more complex than standard anti-fascist notions of fascism as Cesarism or Bonapartism.

Whereas Mussolini saw fascism as a sign of the future, Freud saw it as a symbol of the past. To be sure, Freud, as we have seen, often considered fascism to be rooted in the "Middle Ages." However, fascism equally implied for Freud a return to the issue of primal history, that is the origins of human culture as he had earlier explored the subject in his *Totem and Taboo* and much later in 1930 in his *Civilization and Its Discontents*. In other words, Freud saw fascism as the return of the repressed, more precisely as the primacy of death over life. For Freud, the fight against fascism represented the "eternal struggle between the trends of love and death." Freud was pessimistic, as he wrote in 1931, regarding who would win.[51] For Freud, fascism projected to the political realm the most destructive forces of human desire. Let me explain.

Some months before Freud described Mussolini as a cultural hero, Freud had written a now forgotten text in which he presented Prometheus, the fire bringer, as a *Kulturheros*, the hero of civilization.[52]

Freud presented this myth as a very "obscure" one. Prometheus, according to Freud, had renounced instinctual forces by controlling them. Ironically, this act of control dialectically led to a violent return of instinctual forces of desire. Controlling fire meant, at least in the short term, the possibility of human propagation of destructive fires. In Greek mythology, the heroic actions of Prometheus led Zeus to argue that as punishment for bringing fire to humans, Zeus would make sure that Prometheus and the latter would live forever in misery.

Freud stated "we are aware that the demand for renunciation of instinct, and its enforcement, call forth hostility and aggressive impulses, which only in a later phase of psychical development become transformed into a sense of guilt."[53] In the Greek myth, Prometheus was punished because of his act of defiance against the gods. Civilization did not develop peacefully but rather through an action involving the violent transgression of norms. By not respecting rules (by the act of theft) Prometheus undermined the legitimacy of civilization while also making it possible. Thus, civilization relies on a delicate balance according to Freud. Like Kafka's narratives, civilization is constantly obscuring and revoking itself. Kafka, as Adorno suggests, imagined Prometheus, as finally merging with the rock to which he was chained.[54] In a sense then, Prometheus's worst fate was to be forgotten. Freud tried, by contrast, to remember the hero by unchaining him and recommending him to Mussolini, thus returning the story of Prometheus to conscious political life. Mussolini then became an unbound Prometheus, a fire bringer, or so Freud may have hoped.

Six years before the Mussolini dedication, Freud opposed extreme individualism to civilization and implicitly equated the figure of "dictator" with the persona of a radical narcissist who having seized all the means to power, then rejects civilization and, with it, the Freudian need to renounce instinctual forces.[55] The modern dictator is anachronistic; he signifies the return of the primal father who ruled the "hordes."

Freud was influenced by Karl Abraham's earlier essay on Prometheus which was published in his *Dreams and Myths: A Study in Race*

Psychology (1913). For Abraham the myth of Prometheus was transhistorical, presenting different configurations in different times, nations and cultures. Like Abraham, Freud described the hero as the "creator of man."[56] Prometheus, as a hero, became the metaphor for the primal father that in his essay on "Group Psychology," Freud presented as the original form of human authority. He described this leader as the " 'superman' whom Nietzsche only expected from the future."[57] Unlike the German philosopher, Freud valued the superman for his historical contribution and not as a source of future transcendence. For him, civilization was born, not from the head of the superhero but when the superhero was killed by the group. Only then norms (the law) became detached from the will and fantasies of the hero. In other words, society gave itself the law according to Freud. In a new modern dialectic, the process of democratic will formation which brought emancipation to Jews like Freud, was now threatened from within. Fascism uses democracy to destroy democracy.[58]

By ending the law, the dictator becomes the law and this implies a reversal to the pre-normative epoch of the primal father. This fits Mussolini's rule. In Freud's analogy between Mussolini and Prometheus, Freud may have wished to Mussolini, in implicit anti-fascist terms, the terrible and interminable destiny that the gods ascribed to Prometheus. Both "heroes" shared fantasies of total mastery. But more importantly, Freud, with his usual multi-layered writing style, ascribed Mussolini, the new hero, the innate characteristics of "primitive man" and "primitive ancestors." Later on, in his book on Moses, Freud would describe the hero as an object of political desire, "We know that the great majority of people have a strong need for authority which they can admire, to which they can submit, and which dominates and even ill-treats them. We have learned from the psychology of the individual whence comes this need of the masses. It is the longing for the father that lives in each of us from his childhood days, for the same father whom the hero of legend boasts of having overcome."[59] This line of thought from *Moses and Monotheism* (1939) was written under the spell of a menacing fascist context that eventually led him to exile.[60] In 1933, Austrian dictator Dolfuss dissolved parliament and inaugurated a regime that Freud had called a "moderate fascism." In a letter Freud made clear that he could tolerate Austrian fascism better than "detested" communism. However, with radical fascism, the situation would be different. He wrote to his son Ernest: "Either an Austrian fascism or the swastika. In the latter case, we should have to go."[61]

During the Austrian crisis of 1934, when the Austrian Nazis threatened the life of incipient Austro-fascism and probably referring to Mussolini, Freud had stated: "rumor has it that a certain powerful man insisted on putting an end to the conflict which has been smoldering for long. At some time this was bound to happen." Freud was not happy that a "powerful man" was the guarantor of a dictatorial order. In short, he had to rely on a hero, Mussolini, and his allies, the Austrian fascists. They stood between Freud and his fellow Austrian Jews and the "Nazi scoundrels." It was indeed a double bind that Freud resolved by privately preferring the lesser fascist evil, namely Mussolini and the Austrian fascists. In describing them as: "the heroes and the saviors of sacred order" Freud was indisputably expressing a bitter irony, as he may have been in his 1933 dedication to the Duce.[62]

Freud's condensed irony reflects not only his dependence but also the fascist connection with the primordial hero. For Freud, the myth of the hero is a "lie."[63] The first of a series which includes religion and the mythmaking that went into fascism. Prometheus personifies this mythmaking, lies, theft and other intrigues. Freud described the historical context of global fascism as the dominium of "lies," stealing and deceit.[64] Mussolini as Prometheus became a contextual symptom and in private Freud would define the Duce and Hitler as an "intriguer" and a "thief."[65] According to the Freudian dialectics of life and death, these characteristics follow the principles of destruction. But can fascist destruction lead to self-destruction?

Freud, like many other anti-fascists, wished for an internal self-destructive impulse in which Nazis will fight with each other and kill themselves. But for Freud fascist self-destruction was rooted in the fascist reification of desire, namely a process that often reached a "negative sublime" with respect to the victimization of the abjected Other.[66] Or to put it differently, he saw fascism as a psychotic ideology that in its circular search for full ideological wish fulfillment, it did not consider the external risks that radical fulfillment of desire poses to the ego. In early 1933, in his essay on the question of world vision (*Weltanschauung*), Freud described two visions of the world that shared with religion the status of illusion, "that derives its strength from its readiness to fit in with our instinctual wishful impulses." Freud presented them as "phenomena which, particularly in our days, it is impossible to disregard." These were intellectual anarchism ("a derivate of political anarchism") and communism. There are many reasons to believe that by intellectual anarchism Freud meant fascism. When Freud denounced that the "nihilist" stress on the wishes of the unconscious wrongly

appropriated the theory of relativism for political and aesthetic purposes, he probably had in mind Mussolini's famous presentation of
fascism as political relativism.[67] In Freud's early embracement of the theory of totalitarianism, psychoanalysis was the opposite of religion, and
nihilism the opposite of communism.

I have already explained why Freud was reluctant, at the time he
met with Forzano, to talk about fascism in explicit terms. Only later
would the hidden sources of the metaphor be disclosed. In his *Moses
and Monotheism*, published in his short-lived exile from the fascist powers in London (1938–1939), Freud made explicit connections between
fascism and communism. He nonetheless did not conflate communism
and fascism. He argued that Soviet aims were enlightened and bold but
he criticized Bolshevik means, arguing that the Soviets subjected the
Russian population "to the most cruel coercion and robbed them of
every possibility of freedom of thought." Without noticing the change,
Freud continued, "With similar brutality the Italian people are being
educated to order and a sense of duty. It was a real weight off the heart
to find, in the case of the German, that retrogression into all but pre-
historic barbarism."[68] This retrogression into "pre-historic" time defines
the ultimate distinction between communism and fascism. Fascism represents atavistic forms of desire in politics, the return of the father
whereas communism represents both an idea of the future as well as the
return of the band of brothers that had killed the father.[69] It is highly
symptomatic that in the same book, Freud's description of the workings
of desire is cathected with political metaphors. Let me quote a highly
condensed section:

> All these phenomena, the symptoms as well as the restrictions of per-
> sonality and the lasting changes in character...show a far-reaching
> independence of psychical process that are adapted to the demands
> of the real world and obey the laws of logical thinking. They are not
> influenced by outer reality, or not normally so; they take no notice
> of real things, or the mental equivalents of these, so that they can
> easily come into active opposition to either. They are a state within
> the state, an inaccessible party, useless for the common wealth; yet
> they can succeed in overcoming the other, the so-called normal,
> component and in forcing it into their service.

Freud goes on to describe this loss of rationality as a question of
sovereignty, "If this happens, then the sovereignty of an inner psychical
reality has been established over the reality of the outer world; the way

to insanity is open."[70] The relation with fascism is apparent. In short, Freud considered fascism's irrational quality as mirroring the psychotic detachment of individuals, namely as a collective rejection of reality and as an expression of the death drive. And he thought that self-destruction was a typical outcome for an ideology so deeply rooted in both. He was not wrong in the long term. Hitler and Mussolini finally engaged in a war that destroyed their lives, their regimes and their countries in a final twilight of the self-prescribed heroes, a fascist *Götterdämmerung*. Freud did not live to see this and wrongly identified fascist self-destruction – or the primacy of the death drive that ultimately destroys the life of the ego – with the fascist purges of the 1930s. He, of course, had in mind the Nazi murderous purge of Sturm Abteilung (SA) in 1934 – an event that he recognized having personally enjoyed.[71]

Probably in the mid-1930s Freud could still think that Mussolini presented a moderate fascism. To be sure, Mussolini often called himself a "primitive of the future" but Freud could not risk giving Mussolini the essay on Prometheus the fire bringer that would present the Duce as the primitive of a mythical past, of the primitive boasting hero. Freud knew better. He gave him his discussion on war instead, a subject matter which he argued, should be "a concern for statesmen." In this small book Freud strongly argued that norms were the best means to counteract the violent actions of violent individuals. The Law controls individuals and proscribes dictatorial heroic violence. Freud saw the very existence of solid norms as a barrier against fascism. To be sure, he did not mention that violent individuals or groups could instrumentally use these norms to destroy freedom. Mussolini, a "despot" was implicitly among those rulers that Freud generally criticized in the book as wanting "to go back from a dominion of law to a dominion of violence."[72] Fascism represents this primacy of violence. The message is indeed clear. Violence is politics going backward. Like the forces of desire, fascism the modern form of political desire represents the return of the past.

In Orientalist fashion, Freud does not mention Mussolini in public but refers to the Turks and the Mongols as waging wars that "have brought nothing but evil."[73] Like Voltaire or Diderot, Freud in his public statements used the Orient in order to represent the worst aspects of modern Western civilization. In private he would equate the Nazis with the "Turks" of 1683 when they "were outside Vienna."[74] The image of barbarians at the gate is hardly original but acquires specific contextual connotations. Moreover, Freud defined Mussolini and Lenin as "despots" whom he detested.[75] The idea that "Eastern" despotism is the antithesis of Western culture did not exclude in Freud's mind,

Eastern Judaism. Referring to his brother-in-law he once described him as an Asiatic being in negative terms.[76] In addition he felt the need to explain that Judaism was not "Asiatic" and thereby not fundamentally different to its European "hosts."[77]

Perhaps, there is no better expression of the dialectic of the Enlightenment, of its self-destructive tendencies, than the fact that Freud, the bearer of its legacy, approved its discriminatory and victimizing dimensions. He often described the Jews from the "East" as living images of physical decay and illness. Freud's family, of course, was of Eastern European origin but Freud saw himself as the Westernized counterpoint to despotic Orientalism. According to this "Orientalism" the notion of the East represents the stress on the "death drive", the sum of instinctual forces that "seek to destroy and kill." Freud could not escape some of these destructive dimensions himself. Transference had played a trick on him. But all in all, Freud correctly saw that fascism meant death and violence. It expressed an unbalance in the dialectic between Eros and Thanatos, fusing ethical and aesthetic visual imperatives. In short, fascism represents the return of the repressed. The negative dialectic that Freud never made explicit is that the "repressed" is brought by modernizing forces.

The Orientalist charges notwithstanding, Freud correctly located the return of the repressed as a central part of Western history that had began with Prometheus. If Mussolini was Prometheus, he was then connected to the roots of Western civilization as Freud understood it. This was a much more important dimension in Freud's thought than the Orientalist trends that represented his own negative dialectic. The return of the repressed implies, as in the ambivalent nature of the Prometheus story, a lack of balance between negative and more positive instinctual drives rather their mutual exclusion. As he wrote in the essay he gave to Mussolini, the return of the repressed represented a pendulum between life and death that was radically inclined to death.[78] Freud could not have predicted that Spanish fascist Millan Astray, and later the Romanian fascists, would best personify this lack of balance between Eros and Death when they chanted, "Long live Death (*Viva la muerte*)."[79] This lack of balance, this tension between life and death, was an expression of the fascist need to blur the line between the inside and the outside, that is, between instincts and the external need to repress them.

If for Freud, civilization was born with control, repression and limited denial of death, fascism put forward an ontological rejection of normativity and, at the same time, stressed the overdetermination of death and power. This fascist lack of balance between life and death as previously personified in the deeds of the fire bringer, which is the provider of an element that is simultaneously a metaphor for passionate love and total destruction. Freud saw Prometheus as "a criminal" and a "thief" who had been punished for breaking the norms. In bringing fire to man Prometheus gave a "blow" to instinctual life censoring and limiting instinctual forces. These very instinctual forces that the hero suppressed lived in his inner body, that is, within himself. Tellingly for Freud, Prometheus's phoenix-like liver was the ultimate representative of instinctual forces and even a radical expression of the phallus. Thus, Freud saw the *Kulturheros* as representing both human attempts to control the instinctual drives (what Freud called "the effort to live") as well their reversal in the "death instinct." The *Kulturheros* inaugurated civilization through fire but also "criminally" provided the means for civilization's own instinctual reversal or self-destruction. Freud saw this ambivalence in dialectic terms. This psychoanalytic reading provides a unique insight into Freud's idea of the fascist unconscious. Like Prometheus, fascism for Freud had an ambivalent function. It could bring modernization and, in dialectic fashion, a return of the repressed. Mussolini, the *Kulturheros*, represented modernity and its dialectic outcome: unmediated violence, death and destruction.

This line of argument is central in the psychoanalytically inflected works of the Frankfurt school, especially *The Dialectic of the Enlightenment* (1944) and, perhaps laterally on Wilhelm Reich's *The Mass Psychology of Fascism* published in 1933.[80] In 1939, Melanie Klein described Hitler's weapon as a "destructive and dangerous penis" but perhaps more interestingly she analyzed this weapon as a metaphor of desire, namely of the internal father figure. For Argentine anti-fascist Jorge Luis Borges in 1944, the explanation of fascism was rooted in Freudian notions of the unconscious.[81] Borges was not an exception among critical thinkers. Psychoanalysis, like fascism crossed the Atlantic and traveled the world. Psychoanalysis shaped a broad anti-fascist spectrum that included the Austrian Stefan Zweig; the Italian émigré in Argentina Gino Germani; the Germans Norbert Elias, Erich Fromm and Herbert Marcuse; the African-American thinker Richard Wright; the Greek Cornelius Castoriadis; the Spaniard Rafael Cansinos-Assens; and the Peruvian José Carlos Mariátegui. All of them noted the emancipatory, anti-fascist potential of psychoanalysis.[82] The presence of fascism highly influenced psychoanalysis. It re-shaped the previous psychoanalytic understanding of the dangers of mass politics and their global nature.

Freudian psychoanalysis had a normative, and even moralistic, theory of the unconscious. That is the idea that desire (the id) is potentially negative and should be repressed, controlled and eventually confronted through language and the law. In short, it has to be articulated in rational terms. For Freud, language rather than images, inarticulate feelings or actions, represented the only form of elaboration. Discourse provided the only rational approach to the depths of the unconscious. Other intellectuals, particularly those attracted to fascism did not agree. Georges Bataille, for example, preferred to emphasize the power of fascism's attachment to the inner self of the ego, or what Freud called the unconscious. Bataille, at times became something close to a fellow traveler of fascism and his fascination with its ideology is related to what he perceived as the fascist emphasis on homogenous inner structures.[83] In psychoanalytic terms, this process could be described as the blurring of distinctions between conscious and unconscious forms of desire. Whereas Freud saw fascism as an expression of this "atavistic" mentality, Bataille seemed to dismiss the subjective possibility, the pluralistic heterogeneity, that psychoanalysis put forward.

Fascism, a trans-national ideology could be counteracted with a transnational interpretative engagement. In one of his last writings, Freud presented "some elementary lessons in psychoanalysis," concluding that the recognition of the forces of the unconscious did not imply a rejection of being "conscious." Freud believed that the psychoanalytic stress on the conscious "remains the one light which illuminates our path and leads us through the darkness of mental life."[84] Unlike, Prometheus, and his fascist incarnations, Freud believed that light could, but should not be instrumentalized by heroes because there was the possibility of creating the conditions for the return of the repressed. To put it simply, light as carried by the hero creates the conditions for the emergence of its dialectical opposite: darkness. Like in dreams, the fascist (inevitably) conscious search for the sources of the unconscious allows a process of reversal, what Freud called the transformation of an element into its opposite. This was, of course, the Freudian lesson to Adorno and Horkheimer's *Dialectic of the Enlightenment*, a process that they described as a regression to barbarism, namely as "the enlightenment's relapse into mythology."[85] For Adorno and Horkheimer, myths were obscure and luminous at the same time and they argued that "in fascism ... conscience is being liquidated."[86]

Freud took the first step in understanding the fascist conflation of history, politics and mythology, and to some extent his own resistance to fascism became a symptom of this process. He too seemed to believe in the explanatory power of myths. In the book he gave to Mussolini, he described his approach as "our mythological theory of instincts."[87]

As the anti-fascist Piero Gobetti noted at the time, for Mussolini there was no difference between myths and history, Freud used historical myths to implicitly describe the ideology of fascism.[88]

The daring ironic stance with Mussolini in 1933 that I have analyzed in this essay should not be surprising coming from a master of reading the implicit.

In 1937, for example, condensed irony seemed to be the only response to the fascist stress on the death drive. He had written to Jones: "Our political situation seems to become more and more gloomy. The invasion of the Nazis cannot be checked; the consequences are disastrous for analysis as well. The only hope remaining is that one will not live to see it oneself."[89] The carnivalization of hope works here as a symptom of Freud's often melancholic reading of politics. Prometheus has returned and he is unbound. The past becomes the present, mythical and unmythical. Fascism collapses distinctions. The return of the repressed brings violence, killing and the unchecked control of the forces of nature (the fire the hero stole from the gods). The future is opaque. Freud understands the absence of hope as destruction. As he wrote some years earlier:

A great part of my life's work...has been spent (trying to) destroy illusions of my own and those of mankind. But if this one hope cannot be at least partly realized, if in the course of evolution we don't learn to divert our instincts from destroying our own kind. If we continue to hate one another for minor differences and kill each other for petty gain, if we go on exploiting the great progress made in the control of natural resources for our mutual destruction, what kind of future lies in store for us?[90]

Seeing fascism implies looking at the collapse of rational thinking, namely the negative resolution of the conflict between instinctual drives and the demands of civilization. Political hope is evanescent. But its turning upside down, through ironic condensation, equally represents the analytical gaze that leads to the reassertion of life, even in death.

In a much more dangerous situation, and after the Nazis had annexed Austria with the general consent of its people, Freud had equally addressed the German fascists as symptoms of the reversal of the civilizing process. Before letting the Freuds leave the country, the Gestapo insisted that Freud sign a statement stating the Gestapo had not ill-treated the Freuds. Freud signed adding the comment: "I can most highly recommend the Gestapo to everyone."[91] Freud had a few minutes to think and yet he delivered. Indeed, he told the Nazis, as he had told Mussolini, that they meant destruction.

All in all, it is with critical irony and analytical condensation that Freud thought fascism as the overpowering dominium of desire in politics. Irony, and with it the Freudian drive to think beyond the symptom, escapes many historians in the present as it had escaped fascism in the past. But fascism did not escape the critical gaze of psychoanalysis. It remains the task of interpreters to recognize the anti-fascist potential of psychoanalysis that so many fascists had perceived at the time they were burning Freud's books. The horizon of possibilities opened by this recognition was, and perhaps should still be, a source for a critical self-reflective anti-fascism.

Notes

1. Sigmund Freud, *An Outline of Psychoanalysis* (New York: Norton, 1949) Orig. Pub. 1940, p. 4.
2. For Ernest Jones description of the meeting, see Ernest Jones, *The Life and Work of Sigmund Freud* (New York: Basic Books, 1957) vol. 3, p. 180. Edoardo Weiss had a slightly different version of the events. See Edoardo Weiss, *Sigmund Freud as a Consultant* (New Brunswick: Transaction, 1991) pp. 20–21. Freud presented Weiss as "my friend and pupil." *The Standard Edition of the Complete Psychological Works of Sigmund Freud*, James Strachey, ed. (London: Hogarth Press, 1961) vol. XXI, p. 256.
3. This, of course, was not enough and the *Italian Journal of Psychoanalysis* was banned by the end of 1933, a year of great antisemitic agitation in fascist Italy. One year later Freud thought about the Forzano-Weiss connection regarding the ban. He wrote: "Although Weiss has direct access to Mussolini and received from him a favorable promise, the ban could not be lifted." Letter to Arnold Zweig, September 30, 1934. *The Letters of Sigmund Freud*, Ernst L. Freud, ed. (New York: Basic Books, 1960) pp. 421–422. On the history of Italian psychoanalysis, see Michel David, *La psicoanalisi nella cultura italiana* (Torino: Boringhieri, 1970). On the history Italian fascist antisemitism see Enzo Collotti, *Il fascismo e gli ebrei: le leggi razziali in Italia* (Roma: Laterza, 2003); Michele Sarfatti, *Gli ebrei nell'Italia fascista: vicende, identità, persecuzione* (Torino: Einaudi, 2000); Renzo De Felice, *Storia degli ebrei italiani sotto il fascismo* (Torino: Einaudi, 1993); Meir Michaelis, *Mussolini and the Jews: German, Italian Relations and the Jewish Question in Italy, 1922–1945* (Oxford: Clarendon Press, 1978). On fascist racism, see the excellent study by Aaron Gillette, *Racial theories in fascist Italy* (London: Routledge, 2002).
4. He would later tell Ernst Jones: "Unfortunately the power that has hitherto protected us – Mussolini – now seems to be giving Germany a free hand." Letter to Ernest Jones. March 2, 1937. *The Complete Correspondence of Sigmund Freud and Ernest Jones, 1908–1939*, R. Andrew Paskauskas, ed. (Cambridge: The Belknap Press of Harvard University Press, 1993) p. 757.

5. *The diary of Sigmund Freud, 1929–1939*, Michael Molnar, ed. (New York: Maxwell Macmillan International, 1992) p. 141.
6. "Benito Mussolini mit dem ergebenen Gruss eines alten Mannes der im Machthaber den Kultur Heros erkennt." See the quote in Edoardo Weiss, *Sigmund Freud as a Consultant*, p. 20. See also Ernest Jones, *The Life and Work of Sigmund Freud*, vol. 3, p. 180.
7. Letter to George Sylvester Viereck. July 20, 1928. p. 234. In 1923, however, when he visited Rome Freud did not mention fascism at all. See Letter to Ernest Jones. September 23, 1923. *The Complete Correspondence of Sigmund Freud and Ernest Jones, 1908–1939*, p. 527.
8. Carl E. Schorske, *Fin-de-siecle Vienna: Politics and Culture* (New York: Alfred A. Knopf, 1980).
9. I want to thank Eli Zaretsky for sharing with me his criticism of the image of the "liberal" Freud. The best and most intelligent example of this "liberal" approach is Peter Gay's, *Freud: A Life for Our Time* (New York: W. W. Norton and Co., 1988). For an analysis of the historiography of psychoanalysis, see Michael Steinberg, *Judaism Musical and Unmusical* (Chicago: University of Chicago Press, 2008).
10. Ironically, as Michael Steinberg notes, nothing was more Viennese than Freud's disdain for Vienna. Michael Steinberg, *Judaism Musical and Unmusical*. On how the Austrian context shaped Freud's global understanding, see his "The Catholic Culture of the Austrian Jews" in Michael Steinberg (ed.), *Austria as Theater and Ideology: The Meaning of the Salzburg Festival* (Ithaca: Cornell University Press, 2000).
11. In his subtle analysis of Freud's Moses, the late Edward Said, like his colleague Yosef Yerushalmi before him, seems to be nonetheless engaged in these binaries. As Michael Steinberg argues with respect to Said, he is "less willing to disavow the very idea of identity, the very idea of a boundary." See Michael Steinberg, *Judaism Musical and Unmusical*. See also Yosef Hayim Yerushalmi, *Freud's Moses: Judaism Terminable and Interminable* (New Haven: Yale University Press, 1991) and Edward Said, *Freud and the Non-European* (London: Verso, 2003).
12. Dominick LaCapra, *Writing History, Writing Trauma* (Baltimore: Johns Hopkins University Press, 2000).
13. Letter to Ernest Jones, August 23, 1933. *The Complete Correspondence of Sigmund Freud and Ernest Jones, 1908–1939*, p. 726. Some years later he wrote to Stefan Zweig in 1937: "The immediate future looks grim, for psychoanalysis as well. In any case I am not likely to experiencing anything enjoyable during the weeks and months I may still have to live." Letter to Stefan Zweig. October 17, 1937. *The Letters of Sigmund Freud*, p. 438.
14. On the subject position of exile, see Enzo Traverso, *La pensée dispersée : Figures de l'exil judéo-allemand* (Paris: Léo Scheer, 2004). See also Hannah Arendt's classic section "Between Pariah and Parvenu" in her *The Origins of Totalitarianism* (New York: Meridian Books, 1959) pp. 56–68.
15. Steinberg argues "For the Freud of *Moses and Monotheism*, the survival of subjectivity in exile is enabled by the definition of subjectivity as exile. The reader must then decide whether this position is to be understood as a function of personal exile and old age, in other words as a contingency and a symptom, or whether the position, in its very embrace of its own

contingency and symptomaticity, catches a basic reality of modern subjectivity, enabled by the clairvoyance of Sigmund Freud in contemplation of fascism and its threat to human dignity." Michael Steinberg, *Judaism Musical and Unmusical*, pp. 55–56.

16. On this topic see Eli Zaretsky, *Secrets of the Soul. A Social and Cultural History of Psychoanalysis* (New York: Knopf, 2004) pp. 244–245. For an analysis of the position of the "established" and the "outsider" in modern society, see Norbert Elias, "Notes sur les juifs en tant que participant á une relation établis-marginaux" in *Norbert Elias par lui-même* (Paris: Agora, 1991).

17. Sigmund Freud, *Moses and Monotheism* (New York: Vintage, 1939) Orig. Pub. 1939, pp. 144–145.

18. Here, I appropriate Jacques Lacan's notion of Nazism as a "resurgence" of sacrifice and "a monstrous spell." See Jacques Lacan, *The Four Fundamental Concepts of Psycho-Analysis* (New York: Norton, 1981) p. 275.

19. Slavoj Žižek, *Did Somebody Say Totalitarianism? Five Interventions in the (Mis)use of a Notion* (New York: Verso, 2001) p. 44.

20. For Hanah Arendt's understanding of ideological thinking as the incapacity to think, see Hannah Arendt "Ideology and Terror: A Novel Form of Government," *The Review of Politics*, 15, 3 (1953), pp. 303–327. *The Origins of Totalitarianism*, pp. 158–184.

21. See Eli Zaretsky, *Secrets of the Soul*, p. 245.

22. "My Correspondence with Einstein has been published simultaneously in German, French, and English, but it can be neither advertised nor sold in Germany." Letter to Oskar Pfister. May 28, 1933. *The Letters of Sigmund Freud*, p. 417.

23. See Sigmund Freud, "Why War?" (1932) in his *Collected Papers* (London: The Hogarth Press, 1957) vol. 5, pp. 273–287. For a previous Freudian elaboration of this topic see Sigmund Freud, *Reflections on War and Death* (New York: Moffat, Yard, 1918.) p. 37.

24. See Zeev Sternhell, with Mario Sznajder and Maia Asheri, *The Birth of Fascist Ideology : from Cultural Rebellion to Political Revolution* (Princeton, N.J.: Princeton University Press, 1994) p. 254; Piero Melograni "The Cult of the Duce in Mussolini's Italy," Journal of Contemporary History, 11, 4 (1976) pp. 221–237. As Falasca Zamponi argues, Freud wrote the dedication in a very specific situation. See Simonetta Falasca-Zamponi, *Fascist Spectacle: The Aesthetics of Power in Mussolini's Italy* (Berkeley: University of California Press, 1997) p. 53; Weiss explained the dedication as a result of Freud's putative admiration for the Roman archeological excavations Mussolini had ordered at the time. Forzano, and Peter Gay, seemed to believe that Mussolini, having in mind the dedication, had intervened with Hitler on Freud's behalf, a fact strongly denied by Weiss in his *Sigmund Freud as a Consultant*, pp. 20, 22–21.

25. Eli Zaretsky, *Secrets of the Soul*, pp. 229, 245.

26. See Saul Friedlander, *Nazi Germany and the Jews. The Years of Persecution, 1933–1939* (New York: Harper Perennial, 1998) p. 57.

27. *The Diary of Sigmund Freud, 1929–1939*, p. 149.

28. Raul Hilberg, *The Destruction of the European Jews* (New York: Holmes &Meier, 1985) pp. 5–28. For an analysis of Hilberg, see Federico Finchelstein, "The Holocaust Canon: Rereading Raul Hilberg," *New German Critique*, 96 (2005–2006).

29. Sigmund Freud, *Moses and Monotheism*, p. 117.
30. Letter to Max Eitingon. January 17, 1938. *The Letters of Sigmund Freud*, p. 440.
31. Benito Mussolini, "La dottrina del fascismo" in Benito Mussolini, *Opera Omnia* (Firenze: La Fenice, 1951–1962) vol. XXXIV, pp. 119–121.
32. Ernest Jones, *The Life and Work of Sigmund Freud*, vol. 3, p. 184.
33. See Agnes Heller, *A Theory of Modernity* (Oxford: Blackwell, 1999) pp. 197–199.
34. On the notion of founding trauma, see Dominick LaCapra, *Writing History, Writing Trauma*, p. 81.
35. For my own interpretation of this dimension of fascism, see Federico Finchelstein, "On Fascist Ideology," Constellations (Forthcoming, 2008). See also Enzo Traverso, *The Origins of Nazi Violence* (New York: The New Press, 2003). On the immediacy of the notion of Empire, see Ann Laura Stoler, "Intimidations of Empire. Predicaments of the Tactile and the Unseen" in Ann Laura Stoler (ed.) *Haunted by Empire. Geographies of Intimacy in North American History* (Durham: Duke University Press, 2006).
36. See for example: Letter to Ernst Freud. February 20, 1934, p. 420. *The Letters of Sigmund Freud*: Letters to Ernest Jones. July 23, 1933 and October 15, 1933. *The Complete Correspondence of Sigmund Freud and Ernest Jones, 1908–1939*, pp. 725, 731.
37. "This underarticulation, indeed the suspicion of articulation, functions as well as a critique of ideology. In the context of the baroque regimes of representation, it functions as well as a critique of representation." Michael Steinberg, *Judaism Musical and Unmusical*.
38. See Letter to Ernst Freud. January 17, 1937. *The Letters of Sigmund Freud*, p. 440. Freud also wrote one year later said that Jewish people "owes its tenacity in supporting life; to him, however, also much of the hostility which it has met and is meeting still." He opposed this support of life to "brutality and the inclination to violence." Sigmund Freud, *Moses and Monotheism*, pp. 136, 147.
39. On the concept of musical subjectivity, see Michael Steinberg, *Listening to Reason: Culture, Subjectivity, and 19th-Century Music* (Princeton: Princeton University Press, 2004).
40. On the notion of reactionary modernism, see Jeffrey Herf, *Reactionary Modernism Culture and Politics in Weimar and the Third Reich* (Cambridge: Cambridge University Press, 1984).
41. Sigmund Freud, *Moses and Monotheism*, p. 6. See also Sigmund Freud, *The Psychopathology of Everyday Life* (New York: Norton, 1965) Orig. Pub. 1901, p. 84; Sigmund Freud, *Group Psychology and the Analysis of the Ego* (New York: Norton, 1959) Orig. Pub. 1922, pp. 34–35.
42. Slavoj Žižek, *Did Somebody Say Totalitarianism?* p. 9.
43. Sigmund Freud, *Civilization and Its Discontents* (New York: Norton, 1962) Orig. Pub. 1930, p. 44; Sigmund Freud, *The Question of Lay Analysis* (New York: Anchor, 1964) Orig. Pub. 1927, pp. 42–43; Sigmund Freud, *Introductory Lectures on Psychoanalysis* (New York: Norton, 1962) Orig. Pub. 1917, p. 335. See also Sigmund Freud, "Why War?" p. 287.
44. On the Austrian Catholic context, see Michael Steinberg, "The Catholic Culture of the Austrian Jews." On the Italian Catholic reaction to psychoanalysis see D. Colombo, "Psychoanalysis and the Catholic Church in Italy:

the Role of Father Agostino Gemelli, 1925–1953," *Journal of the History of the Behavioral Sciences*, 39, 4 (2003) pp. 333–348. See also See Michel David *La psicoanalisi nella cultura italiana* (Torino: Boringhieri, 1970).

45. See Sigmund Freud, "Why War?" pp. 284, 287.
46. Gramsci was himself, often reticent about the theoretical possibilities opened by psychoanalysis, rather "psychoanalytically" he presented Freud's theories as a super structural symptom of modern capitalist society in 1928. In 1931, Gramsci argued that Freud like Lombroso wanted to build a "general philosophy" out of mere empirical observations. Later in 1935, Gramsci believed that psychoanalysis was a "science" that worked better with the upper classes and was not necessarily fit to study the subaltern classes. Nevertheless, Grasmci was highly receptive of Freud's interpretation of dreams and related the Freudian analysis of the Oedipus complex to a "new revolutionary ethics." See Antonio Gramsci, *Letteratura e vita nazionale* (Roma: Editori Riuniti, 1977); Antonio Gramsci, *Gli Intelletuali* (Roma: Editori Riuniti, 1977) p. 94; Antonio Gramsci, *Passato e presente* (Roma: Editori Riuniti, 1977) pp. 284–285. In private matters, however, he was much more open to psychoanalysis, including his hopeful support of his wife's psychoanalytic treatment. See Aurelio Lepre, *Il Prigionero. Vita di Antonio Gramsci* (Roma-Bari: Laterza, 2000) pp. 148–149; Antonio Gramsci, *Letters from Prison* (New York: Columbia University Press, 1994) vol. 2, p. 29. From his fascist jail Gramsci admitted: "Non ho potuto studiare le teorie di Freud." The fact that Gramsci was a victim of fascism, eventually promoting his death left us without this needed study.
47. Sigmund Freud, *Civilization and Its Discontents*, p. 41.
48. I borrow here from Louis Althusser's reading of the political implications of psychoanalysis, Louis Althusser, *Machiavelli and Us* (London: Verso, 1999) pp. 117, 126. On the uncanny and anti-fascism, see Eli Zaretsky's chapter in this book: "Beyond the Blues: Richard Wright, Psychoanalysis, and the Modern Idea of Culture."
49. See Giovacchino Forzano (with Benito Mussolini), *Campo di Maggio dramma in tre atti* (Firenze: G. Barbèra, 1931). See also Giovacchino Forzano, *Mussolini; autore drammatico con facsimili di autografi inediti* (Firenze: G. Barbèra, 1954).
50. Sigmund Freud, *Group Psychology and the Analysis of the Ego*, pp. 34–35.
51. Sigmund Freud, *Civilization and Its Discontents*, pp. 96, 112.
52. In his path-breaking analysis of this myth, Italian psychoanalist G. Contri argues that Freud's encrypted message to Mussolini was a sarcastic insult, namely that he was a thief. Contri's approach is symptomatic of the old anti-fascist cliché that fascism was above all corrupt and thus it did not warrant a more profound analytical investigation. He argues that Freud had no ambivalence toward Prometheus. My reading is, of course, different and stresses the relation between Freud's characterization of Mussolini as Prometheus and the myth's relation to both the origins and the breaking down of culture. In arguing against the Antifascist view that fascism is just barbarism, I connect the idea of barbarism to Freud's understanding of the unconscious nature of the limits to civilization as posed by the return of the repressed. See Giacomo B. Contri, *Lavoro dell'inconscio e lavoro psicoanalitico* (Milano: Sipiel, 1985) pp. 90–93.
53. Sigmund Freud, "The Acquisition of Power over Fire" (1932) in his *Collected Papers* (London: The Hogarth Press, 1957) vol.5, pp. 290–291.

54. Theodor Adorno, *Prisms* (Cambridge: MIT press, 1983) p. 247.
55. See Sigmund Freud, *The Future of an Illusion* (New York: Norton, 1961) Orig. Pub. 1927, pp. 6, 15. In this regard, Julia Kristeva argues: "The more or less beautiful image in which I behold or recognize myself rests upon an abjection that sunders it as soon as repression, the constant watchman, is relaxed." See Julia Kristeva, *Powers of Horror: An Essay on Abjection* (New York: Columbia University Press, 1982) p. 13. For Kristeva abjection and abjecting represent a "precondition of narcissism."
56. See Sigmund Freud, *The "Wolfman" and Other Cases* (London: Penguin, 2003) p. 83. See also Karl Abraham, "Analysis of the Prometheus Saga," *Dreams and Myths: A Study in Race Psychology* (New York: The Journal of Nervous and Mental Diseases Publishing Company, 1913) p. 27.
57. Sigmund Freud, *Group Psychology and the Analysis of the Ego*, p. 71. On the hero as appearing in history between the totem and the God, see Sigmund Freud, *Moses and Monotheism*, p. 171.
58. As Gillian Rose perceptively notes, fascism abolishes the distinction between fantasy and political action. See Gillian Rose, *Mourning Becomes the Law. Philosophy and Representation* (Cambridge: Cambridge University Press, 1996) p. 53.
59. Sigmund Freud, *Moses and Monotheism*, pp. 139–140.
60. For Michael Steinberg: "The text's trajectory is complicated. Freud begun work on it in Vienna in 1934, published its first two sections in *Imago* in 1937, and continued to work on it in exile in London in the second half of 1938. The German text was published in Amsterdam in 1939." Steinberg argues that the book presents two subject positions, namely the pre-exile and post-exile voices.

The pre-exile argument claims to take Moses away from the Jews. It destabilizes Jewish historical paternity. The post-exile voice claims the position of a new father. In exile, Freud is Moses – a new Moses. Here is the problem: does exile work through the problem of an excessive identification with the super-ego, with the replacement of the father? In other words, does Freud take on the voice of Moses only from the vantage point of exile, so that Freud's Moses – Freud as Moses – assumes the mantle of the father without fulfilling the mortal contract between primal father and son? Replacing the father in the primal contract requires patricide. Exile, however, may literally shift the grounds of succession adequately to provide an escape from the primal contract and from the bounds of patricide.

See Michael Steinberg, *Judaism Musical and Unmusical*, Chapter 2. See also Eric Santner, *On the Psychotheology of Everyday Life. Reflections on Freud and Rosenzweig* (Chicago: University of Chicago Press, 2001) p. 7.
61. Peter Gay, *Freud: A Life for Our Time*, p. 595.
62. Letter to Ernst Freud. February 20, 1934. *The Letters of Sigmund Freud*, p. 419.
63. Sigmund Freud, *Group Psychology and the Analysis of the Ego*, pp. 88–89.
64. "The political situation...It seems to me that not even in the War did lies and empty phrases dominate the scenes as they do know." Freud had the

opportunity to see Nazi corruption at his home, when a gang from the Sturm Abteilung had forced their way into the dining room. Mrs. Freud fetching the household money, she put it at the table ironically telling the Nazis: "Won't the gentlemen help themselves?" In addition, Anna Freud gave them the money from the safe. They took $840 and Freud later observed that "he had never been paid so much for a single visit." Ernest Jones, *The Life and Work of Sigmund Freud*, vol. 3, pp. 181, 219.

65. Freud told Jones in 1934: "Perhaps at this very moment the intriguer M. in Venice is selling us to the captain of the thieves H." Letter to Ernest Jones. June 16, 1934. *The Complete Correspondence of Sigmund Freud and Ernest Jones, 1908–1939*, p. 737.

66. On the notion of negative sublime and fascism, see Dominick LaCapra, *History and Memory after Auschwitz* (Ithaca: Cornell University Press, 1998) pp. 27–30; *Representing the Holocaust. History, Theory, Trauma* (Ithaca: Cornell University Press, 1994) pp. 100–110 and *Writing History, Writing Trauma*, p. 94.

67. Mussolini opposed fascist relativism to "scientificism." See Benito Mussolini, *Diuturna* (Milano: Imperia, 1924) p. 375. Many anti-fascists such a Piero Gobetti, noted the centrality of Mussolini's relativism. See Piero Gobetti, "Benito Mussolini" in his *On Liberal Revolution* (New Haven: Yale University Press, 2000) p. 58. It is interesting to note that Hannah Arendt was equally interested in this aspect of Mussolini's ideology and connected it with the romantic tradition. Hannah Arendt, *The Origins of Totalitarianism*, p. 168. Sigmund Freud, *New Introductory Lectures on Psychoanalysis* (New York: Norton, 1965) Orig. Pub. 1933, pp. 175–176.

68. Sigmund Freud, *Moses and Monotheism*, p. 67.

69. I want to thank Eli Zaretsky for sharing with me his interpretation of Freud's notion of communism as the historical return of the band of brothers that had killed the father.

70. Sigmund Freud, *Moses and Monotheism*, p. 96.

71. "(Freud) told Arnold Zweig that his impression of the event presented a striking contrast to an experience he had had at the Hague Congress in 1920. There the hospitable Dutch had invited their half-starved colleagues from central Europe to a sumptuous banquet, being used to so little food they found the hors d'oeuvres a sufficient meal and could not eat more." Reflecting on this Freud ironically argued" "Now the hors d'oeuvres in Germany leaves one hungry for more." Ernest Jones, *The Life and Work of Sigmund Freud*, vol. 3, p. 189.

72. See Sigmund Freud, "Why War?" p. 277.

73. Ibid, p. 278.

74. Letter to Ernest Jones. March 2, 1937. *The Complete Correspondence of Sigmund Freud and Ernest Jones, 1908–1939*, p. 757.

75. Letter to George Sylvester Viereck. July 20, 1928. *The Letters of Sigmund Freud*, p. 381.

76. Sander Gilman, *Love + Marriage = Death and Other Essays on Representing Difference* (Sanford: Stanford University Press, 1998) p. 58.

77. In the same paragraph, and without noticing the change, Freud goes on to describe the fundamental aspect of Judaism, namely its constitutive challenge to oppression, Sigmund Freud, *Moses and Monotheism*, pp. 115–116.

See also on Freud's relation vis-à-vis Eastern Jews, Sander Gilman, *Freud, Race and Gender* (Princeton: Princeton University Press, 1993).

78. Freud argues "the gratification of these destructive impulses is of course facilitated by their admixture with others of an erotic and idealistic kind." See Sigmund Freud, "Why War?" p. 282.

79. On the fascist theory of the abject, see Federico Finchelstein, "On Fascist Ideology."

80. Eli Zaretsky, *Secrets of the Soul*, pp. 217–249, 264. In the case of the Frankfurt School, it may be argued that the all-encompassing "Dialectic of Enlightenment" was triggered by the effects that fascism was at the time imposing upon modernity. Max Horkheimer and Theodor Adorno, the authors of the book, even argued that fascism seemed to be a phenomenon that would not vanish with the destruction of the Hitler or Mussolini regimes.

81. Borges, of course, had an ambivalent, and sometimes ironic, relation vis-à-vis psychoanalysis. For Borges's understanding of fascism through Freud, see "Anotación al 23 de Agosto de 1944," Otras Inquisiciones, *Obras Completas* (Buenos Aires: Emecé, 1996) T.2, p. 105. Borges later explored the ideological dimensions of fascism in his story Deutsches Requiem. For an analysis of Borges in this context, see Federico Finchelstein, "Borges, la Shoah y el 'Mensaje kafkiano'. Un ensayo de interpretación." *Espacios de Crítica y Producción. Publicación de la Facultad de Filosofía y Letras-Universidad de Buenos Aires*, 25 (1999) pp. 75–80.

82. See Theodor Adorno, "Freudian Theory and the Pattern of Fascist Propaganda" in Andrew Arato and Eike Gebhardt (eds.), *The Essential Frankfurt School Reader* (Oxford: Blackwell, 1978) pp. 118–137; Renzo De Felice, *Interpretations of Fascism* (Cambrigde, Mass: Harvard University Press, 1977) pp. 78–87; Martin Jay, *The Dialectical Imagination* (Berkeley, University of California Press, 1996) pp. 87–112; Enzo Traverso, *Understanding the Nazi Genocide. Marxism alter Auschwitz* (London: Pluto Press, 1999). Cornelius Castoriadis, *Figuras de lo pensable* (Valencia: Universitat de Valencia, 1999) pp. 179–192. See also Eli Zaretsky, "Beyond the Blues: Richard Wright, Psychoanalysis, and the Modern Idea of Culture"; Alejandro Blanco, *Razón y modernidad: Gino Germani y la sociología en la Argentina* (Buenos Aires: Siglo XXI, 2006).

83. See Georges Bataille, *Visions of Excess* (Minneapolis: University of Minnesota Pres, 1985) pp. 137–160.

84. Sigmund Freud, "Some Elementary Lessons in Psycho-Analysis" in his *Collected Papers* (London: The Hogarth Press, 1957) vol. 5, p. 382.

85. See Max Horkheimer and Theodor W. Adorno, *Dialectic of Enlightenment* (Stanford: Stanford University Press, 2002) p. 164.

86. Ibid, p. 164.

87. See Sigmund Freud, "Why War?" p. 283.

88. See Piero Gobetti, "Benito Mussolini" p. 58.

89. Letter to Ernest Jones. March 2, 1937. *The Complete Correspondence of Sigmund Freud and Ernest Jones, 1908–1939*, p. 757.

90. Letter to Romain Rolland. March 4, 1923. *The Letters of Sigmund Freud*, pp. 341–342.

91. Peter Gay, *Freud: A Life for Our Time*, p. 628; Ernest Jones, *The Life and Work of Sigmund Freud*, vol. 3, p. 226.

Section 3
The Transnational Diffusion of Psychoanalysis

5
The Travelling Psychoanalyst: Andrew Peto and Transnational Explorations of Psychoanalysis in Budapest, Sydney and New York

Joy Damousi

> *We are grateful that Andrew Peto – having voyaged from Middle Europe to the Antipodes, saw fit to uproot himself and travel again – half way around the world – to settle permanently in New York.*[1]

When the Hungarian born psychoanalyst, Andrew Peto, rose in August 1951 to address a small audience who had assembled in a room at British Medical Association House in Macquarie Street, Sydney, it is true to say that a historic moment had arrived. Peto's inaugural paper on the prevention of juvenile delinquency, to the newly constituted Sydney Institute of Psychoanalysis, christened the birth of a new institution. The institute was barely two-months old, having been formed by Peto, the leading Sydney psychoanalyst Roy Winn and Siegfried Fink. Fink, like Peto, was a European immigrant who had arrived in 1938 from Germany; he worked in the health services to have his medical qualifications recognised, and then became a practicing analyst in Sydney.

It was not only the birth of a newly formed Institute that made this moment historic, significant enough as it was for a newly formed psychoanalytic institute to be established. Peto's paper and his presence in Australia signified two further points of historic interest. It first of all pointed to an under recognised, yet defining characteristic of the history of psychoanalysis, that is, its migratory, transnational aspect, which involved the movement of analysts and psychoanalytic ideas across many continents, especially after World War Two. Second, Peto's presence in Australia between 1949 and 1955, and then his migration to New York from 1956 where he remained until his death in 1985, raises the suggestion of the borderless state of psychoanalysis as a theory and a practice. As a part of the Hungarian diaspora, Peto brought with him in his travels a perspective which was defined and shaped by his studies in Hungary – especially that of the work of Michael Balint and Sandor Ferenczi – but he applied the interpretation of Freud offered by these practitioners transnationally and cross-culturally.

This chapter attempts to cast Peto's peripatetic movements within a framework that takes as its underling premise the view that psychoanalysis is transnational, by its very theoretical structure. Although deeply embedded within a white, Eurocentric, Western paradigm, psychoanalysis is a theory of the unconscious that can be applied across cultures – an aspect which Freud's disciples in anthropology in particular have applied in their own psychoanalytic work with great enthusiasm.[2] And yet paradoxically, it is evident that certain psychoanalytic theories take root in nations and cultures and others do not. The question of whether a distinctive psychoanalytic approach is culturally specific has preoccupied scholars of Freud and his theories, and most of the histories written thus far of this process have been analysed within a national framework.[3]

In this chapter, I wish to move beyond this national paradigm, and to pursue instead the paradox of analysing the ways in which psychoanalysis can work at both and at once at the level of a national and transnational theory and practice. I argue that national and international discourses are inter-related rather than distinctive processes operating independently and separately. For the travelling psychoanalyst, an examination of issues such as juvenile delinquency, child's play, countertransference, dream analysis – to take a few topics of Peto's writings – transcended national borders but were shaped in a specific national context. Peto's sojourn in Australia and his subsequent travel to New York reflect these intersections, as his psychoanalytic education was shaped and formed by a distinctive analytic training in Hungary, but was seamlessly transposed during the 1950s, to psychoanalytic cultures and worlds as different as Sydney and New York. Peto's medical training allowed him to move eclectically within these worlds, implying that the unconscious was transnational, and in clinical practice, there was no hierarchy of ideas which reflected a centre and a periphery relationship. Instead, the 'centre' and 'periphery' merged as one wherever he happened to find himself.

*   *   *   *   *

Andrew Peto was born in Budapest in November 1904. After pursuing medical studies in Switzerland, Austria and Hungary, he graduated from the University of Budapest in 1929. With an interest in pediatrics, he worked at the pediatric clinic at the University of Greifswald between 1930 and 1932; he then moved to Budapest and worked at the Jewish Children's Hospital from 1932 to 1937. In 1935, he began training as a psychoanalyst at the Hungarian Psychoanalytic Society, graduating in 1938. Between 1938 and 1949, Peto practised psychoanalysis in Budapest. After the war, he was the psychotherapist in charge in the Public Health Service from 1945 to 1949; between 1948 and 1949, he was also psychiatrist to the juvenile court in Budapest. After his arrival in Australia in 1949, he became a training analyst at the Sydney Institute of Psychoanalysis and one of its directors. He taught and practiced psychoanalysis in Sydney from 1950 to 1956. In 1956, Peto moved to New York, becoming clinical Professor of the Department of Psychiatry at the Albert Einstein College of Medicine. He taught there until he resigned in 1974.[4] Peto became a member of the New York Psychoanalytic Society in 1958; a training analyst in 1962; and President of the Society from 1975 to 1977.

For Peto's generation of analysts, such migration and movement across the world was not unusual. Melanie Klein, for example, who was born in Vienna, travelled to Budapest to undertake analysis with Ferenczi, and then moved to Berlin in 1921 where she was analysed by Karl Abraham. In 1926, she was invited to London by Ernest Jones, where she worked and lived until her death in 1960. Karen Horney worked at the Institute of Psychoanalysis in Berlin from 1920. In 1930, she migrated to the United States where she joined a lively intellectual community that included Eric Fromm and Harry Stack Sullivan. The war of course created circumstances that forced analysts to seek a professional future outside of Europe. Erik Erikson, a graduate of the Vienna Psychoanalytic Institute in 1933, emigrated to Denmark, then to the United States following the Nazi rise to power. Helene Deutsch fled Germany in 1935, reluctantly migrating to the United States. Freud himself succumbed to the imminent dangers of Nazi rule and left Vienna in June 1938 for London.

Within this context of international movement, can Peto's activities and writings be considered transnational? Sven Beckert argues that 'transnationalism' can be understood 'as an approach to history that focuses on a... range of connections that transcend politically bounded territories and connect various parts of the world to one another. Networks, institutions, ideas and processes constitute these connections, and though rulers, empires and states are important in structuring them, they transcend politically bounded territories'.[5] Historical processes are constructed in such accounts, according to Isabel Hofmeyr, at the very 'movement between places, sites, and regions'.[6] Ann Curthoys and Marilyn Lake identify transnationalism as a set of 'processes and relationships that have transcended the borders of nation states', that offer 'insight into interconnectedness of political movements and ideas'.[7] What does Peto's movements and treatments tell us about the treatment of the unconscious when transcending national borders in the ways in which Beckert and Hofmeyr suggest?

An analyst such as Peto applied Freudian ideas and theories across different contexts, places and times. In his own reflections and writings, such applications are cross-cultural and transnational. Do experiences in Hungary, Australia and America produce different psychological experiences that can be culturally and nationally defined? In his study of cultural pluralism and psychoanalysis, Alan Roland has argued that while psychoanalysis has derived from the Western and philosophical tradition, its tenets have been applied to a range of cultures with different historical traditions, such as those in Indian and Japanese cultures. He concludes in this study that applying such an 'individualistic psychology' to specific cultural contexts, such as psychoanalysis, is problematic as it creates reductionist and often crude generalisations about specific cultures and the unconscious.[8] And yet, as a practicing psychoanalyst in New York, 'who has worked with a variety of American patients from different ethnic groups, social classes, and regions of the United States, as well as with Indians and Japanese in their own countries and in New York City', he has been made aware of the need to consider cultural and social factors, in order to resolve inner conflicts.[9]

In contrast, Peto adopted a transnational approach in his writings that was medical in orientation, taking his particular from of psychoanalysis to his patients wherever they were to be found. As an analyst trained in a period when theoretical applications of psychoanalysis were universal, Peto took his approach beyond national boundaries, which also transcended cultural considerations.

The Hungarian Psychoanalytic Society

Peto spent his formative years as a member of the Hungarian Psychoanalytic Society. This was a multi-disciplinary, eclectic society that reached the height of its popularity after the First World War. Peto was not the only member of the society to forge connections with the Antipodes. One of its most prominent members, the anthropologist Geza Roheim, travelled to Australia and made a major contribution to psychoanalytic ideas through his discussion of 'collective trauma' during the course of his studies in that country. He believed that the impact of these traumas could be deciphered in the myths and rituals of a given culture. Another member of the Society provided a more enduring connection. Clara Lazar-Geroe, migrated to Australia in 1940 and established the first institute of psychoanalysis in Melbourne. Lazar-Geroe recalled that there was a very dynamic psychoanalytic community in Hungary immediately after World War I. Termed the 'Budapest School', the Society was driven and shaped by Sandor Ferenczi, one of Europe's leading psychoanalysts and theorists during the inter-war years. Known for his interdisciplinary approach and his flexibility – and those who followed him – created a centre of psychoanalysis that rivalled Vienna for its intellectual and avant-garde culture. Ferenczi's intellectual engagements and innovations created a distinctive psychoanalytical technique that involved an emphasis on child therapy and the early mother–child relationship, and laid stress on the influence of countertransference.[10] Ferenczi gained a reputation for his work on child studies, and his impact can be gauged in Australia through Lazar-Geroe and her subsequent training of other analysts here.

Radically, Ferenczi had shifted analysis away from the Oedipus complex in favour of the separation of mother and child, and linked infantile sexuality to that of the mother–infant relationship. Michael and Alice Balint, who were also a part of the Budapest group, discussed the relationship between the mother and the infant. In 1937, Balint identified the distinctive element of the Budapest School as, 'the importance of the psychological development in the first years of life'. Ferenczi also pioneered the need to 'frustrate transferential desires' and not to remain neutral as an analyst, as Freud has suggested. It was only when the analyst was 'put on the patient's level of repression' that he or she could be understood. Balint extended this examination by considering the relationship of the analyst and analysand as a mutual relationship, 'in which transference and counter transference are intertwined'.[11]

As the Hungarian Psychoanalytic Society had historically taken a strong position in favour of non-medical analysis, it was therefore able to accommodate therapists such as Ferenczi, who promoted a more interactive method with his patients and supported a 'greater liberty and a freer expression in behaviour of their aggressive feelings towards their physicians'. But this did not mean that the demand for *training* before the term 'analyst' could be used was compromised – either in Hungary or Australia.

The circumstances which led to Peto becoming a world citizen were not of his own choosing as was the case with other analysts who were forced to flee Europe before, during and after the period of Nazi occupation. Peto was fortunate to escape the fate of so many of his fellow Hungarians. The Nazi forces annihilated the Jewish population in Hungary almost completely. Hungary entered the war in June 1941, and by 1944 it was occupied by German forces. Thereafter there was a systematic destruction of Jewry in Hungary, with the Nazis annihilating the political, legal and social framework of Jewish existence.[12] For Peto, this also involved a personal tragedy. Peto's first wife, Dr. Elizabeth Kardos, a child analyst, was killed by Nazis during the siege of Budapest in January 1945. Prior to the war, Kardos had been one of the analysts who attempted to escape Europe, but was unable to do so. Her name was listed among the analysts who were trying to enter Australia.[13]

Judit Meszaros has traced the migration of Hungarian analysts throughout the world following Nazism. Meszaros documents how a vibrant and thriving psychoanalytic community emerged after the First World War in Budapest, the inter-war years saw the destruction of this community after 1933. Andrew Peto had attempted to come to Australia to flee Nazism in 1938. Peto was named as part of a group who were trying to migrate to Australia. Geroe, Peto, Elizabeth Kardos and Eva Rosenfield were agitating for migration in March 1939.[14] Meszaros states that Kardos and Peto were apparently granted visas but did not take up the offer of migrating to Australia.[15]

In an interview conducted in 1984, a year before his death, Peto discusses the atmosphere of war-time Budapest. In a climate of heightened suspicion, paranoia and mistrust, Peto continued to practice analysis as best he could under the circumstances:

> of course, many patients stopped coming because they were preoccupied, how to save their lives, and what to do, and where to go, or where to stay. But still I had some practice even during the war there was no problem except those who were killed or taken [to] labor camps . . . I stopped seeing patients when the getto (sic) arrangements came . . . [16]

For Peto, psychoanalysis was not a political theory or one which was designed to account for social, political and cultural circumstances. Psychoanalysis was

Conceived to solve certain emotional problems. But it was not conceived to cope with cataclysmic events in the environment. It was
conceived as a therapy, and then a theory, but neither the therapy
nor the theory were put into the frame of historical events.[17]

When asked if the war had an impact on his analytic thinking,
he resisted the invitation to consider what the impact would be;
indeed, 'For an analyst there's nothing surprising about war'.[18] Of his
analysands, he stated, it was reasons associated with their private life
that they came to see him, not the political situation. But he resisted
making a connection between the two aspects.[19]

After the war, immigration to Australia provided expanded opportuni-
ties for post-war immigrants. Peto's arrival in Australia in 1949 coincided
with the largest immigration programme undertaken by Australia. Following the war, there was a continuing stream of Hungarian refugees,
many of whom were Jewish. In 1949, the year Peto arrived, 3553 men
and 2078 women arrived in Australia. One of the conditions of arrival
was that immigrants sign a two year contract on employment with
the Australian government. Although immigrants' qualifications were
not recognised, Peto appears to have had no trouble being included in
the psychoanalytic fraternity.[20] By June 1954 there were about 16,000
Hungarians in Australia. These numbers would soon swell by another
14,000 after the 1956 Hungarian revolt. Two of the later Hungarian
arrivals included Peto and Vera Roboz, who arrived soon after the 1956
invasion of Hungary by the Soviet Union.

Peto crossed the border to a foreign and alien culture. Sydney in
the 1950s did not have the vibrant European flavour and cultural cosmopolitanism of Budapest. But it did provide him with a material
and psychological haven from the ruins of war. Above all, it provided
him with an opportunity to continue to practice unencumbered as an
analyst.

Sydney Institute of Psychoanalysis

On his arrival in Sydney, Peto would have found a small, but dedicated psychoanalytic community. The first psychoanalytic institute
in Australia, the Melbourne Institute of Psychoanalysis, formed in
Melbourne in 1940, had been building and developing a psychoanalytic practice and training in Australia for ten years. The advent of the
Sydney institute in 1951 was a welcome addition to these endeavours to
establish training institutes in Australia. While the formal institutional

structures of psychoanalysis were in its infancy, Freudian ideas have been in circulation in a range of fields. Throughout the first half of the twentieth century, such ideas had been in currency in the universities, the child welfare profession, social work, in education circles and in the medical profession. Beyond these areas, Freudian theories found currency among artists and in bohemian cultures.[21]

Along with Roy Winn – whose endowment helped to establish the institute – Peto and Siegfried Fink became its directors. Interstate directors of the Sydney Institute included Clara Lazar-Geroe. Peto was the correspondent for the Institute and reported in its events and activities to the British Psychoanalytic Society.[22]

Peto published articles on several themes, but the two which are especially dominant in his writings were those of children and the transference relationship – both themes which were central to Ferenczi's theories and preoccupations.

Peto had a particular interest in delinquency. His experience as a psychiatrist with the Juvenile Court in Budapest, allowed him the 'opportunity to examine thoroughly about 150 juveniles, most of whom were lads aged between twelve and eighteen years'. What were the defining aspects of juvenile delinquency? He viewed the issues to be associated to juvenile delinquents – as they were called in the 1950s – related to the earliest childhood of juvenile delinquents – the first three to five years of life, and particularly the child–mother relationship was of paramount importance. He defined two categories: 'that of the extremely neglected children, and that of the extremely spoiled children'. A 'lack of care and love caused permanent emotional frustration which disturbed the child's attachment to the parents and to the siblings'. In order to illustrate the importance of 'psychological trauma in early infancy as a cause of delinquency':

A patient who had spent years in the prostitute underworld, where he lived a bisexual life, had been extremely neglected and ridiculed by both father and mother. When three years old he had set fire to their flat. Later on he had regularly stolen money from his parents. He had married a prostitute, and been imprisoned for attempted blackmail when he caught a man with his wife in a prearranged sexual situation. His childhood experience was one of frustration and neglect.[23]

There was other category of criminals who were labelled 'spoiled children' a condition which could be traced to their mothers. What did these mothers do?

When young they received excessive attention from over-anxious parents, mostly mothers, and the first years of life were spent in very close emotional contact with their parents.[24]

Peto concluded that 'practically all delinquent children were neurotic or at least showed definite emotional disturbances'. It was the case he believed, that most cases could be traced back to 'early traumata'.

Given his own personal experiences, the psychological impact of war was a key theme in Peto's writing which also included its impact on children. He argued that

during and after the siege of Budapest prolonged hunger and under-nourishment combined with continued fear of death, general instability of living conditions, and pronounced disorganisation in personal and civic hygiene caused deep changes in personality in those particularly who had previously lived under safe and established conditions.

These conditions persisted in children, even though the majority did not become delinquent. As adults, the starvation and deprivation – which often included stealing food – in war revealed anxieties of earliest childhood. To act 'delinquently' meant an attempt to overcome that kind of anxiety.[25] Peto's talk generated much interest and discussion from the audience of analyst who attended his paper and in response to queries, believed that while the majority of children 'exposed to difficult social conditions did not become delinquent', it was 'whether a child was loved or not was more important'. Boys were more easier for him to treat, and he stressed 'over-indulgent parents', and 'lack of family standards' which created anxiety.[26]

Childhood was a key concern and theoretically it was an area of study which Peto systematically examined throughout his professional life. In August 1953, the World Health Organisation conducted an international seminar on 'Mental Health in Childhood' that aimed to draw together workers in the fields of child health, education, sociology and child welfare internationally, to consider factors which affect the growth and development of children from birth to six years. Peto was part of a discussion on the effects of separation on children, a large and significant theme in the conference. Issues raised in this discussion included questions such as why some children could tolerate separation better than others in apparently similar situations; separation as a socio-medical problem seemed to be greater in Western societies with their smaller, more isolated families. The war in particular had focussed the attention and mind on the separation of children:

> The large-scale military operations of the second World War had given a sad opportunity to observe the devastating effect on the mind of the growing child. Relatively the slightest trauma was that represented by the father's absence. Love as well as hate were concentrated on the mother. Ambivalence developed to a higher degree. The sense of guilt could not develop evenly, because one of the main figures – the source of helpful identifications – was missing.[27]

There were other issues which developed in this context. Children, it was observed, blamed their mother for their loss, in relation to the father's absence: 'Lack of family security increased in the anxiety situations and aroused an excess of aggression in many children'.[28] Several other factors were identified as counteracting the disruption of the home. These included: a well-balanced mother gave the child 'a feeling of security. The mother's thoughtfully protective love was a strong barrier between the child and the outer world; the child's opportunity to play was threatened if the danger situation heightened; a constant environment was of paramount importance and change was disruptive; traumatic situations were aggravated by starvation as the starving child blamed its mother who became representative of a neglectful world'.[29]

The influence of child analysts such as Ferenczi is evident early in his work, especially with his wife, Elizabeth Kardos. In a paper on the dynamics of play and trauma, Peto draws extensively from theories of Ferenczi. Play, he argues, 'endeavours to annul the trauma' through 'integrating the split-off complexes into the child's personality. Simultaneously an attempt is made to create an atmosphere of blissful gratification instead of the frustrating trauma'.[30] These arguments are explored in relation to the body, the nipple and the child's play of the mother's body. Peto brought an international perspective to the Sydney institute as it was developing; his training was shaped by Ferenczi and he brought those interests to his Antipodean position.

It was timely that one first presentations to the institute would be given by Peto on shell shock. Arriving from war-ravaged Europe, Peto spoke on the ways on which the symptoms of war neurosis had shifted from World War I to World War II. The main difference he discerned was that there had been a change from the impact on muscular, motor actions of shell shock, to 'depression, combat fatigue, working disability and radical changes of the character'.[31] In this discussion he again was especially interested in children, and following Ferenczi, in the relationship between the child and the adult, especially in relation to the physical and mental differences between adult and child; the ego and superego of the child.

His involvement with war veterans was conducted through an examination of 62 ex-soldiers with chronic war neurosis. In his summary of the findings in this examination, he saw it as an opportunity to discuss traumatic experiences of early childhood. In his conclusions, Peto made several observations in relation to the treatment of children with adults such as maltreatment by parents or foster-parents; 'emotional and material privations of all kinds'. There was enduring and long-lasting hatred of their parents or their representatives, 'conscious from early childhood', and a sense of guilt. What were the symptoms of this condition? These were: 'depression, emotional indifference towards relatives...fits of anger finding outlet in beating of the relatives, and varying degrees of impotence'.

Peto contextualised these developments in terms of wider issues related to the war. This increase in guilt feeling was also precipitated by the way in which modern warfare was fought. The role of fantasies was also central; the experience of war 'revived hostility against [parents] aroused and strengthened the old guilt feelings of early childhood'.[32]

In this period, Peto also published at length on the question of transference and countertransference relationship in psychoanalytical therapy. Drawing on classical Freudian analysis of transference, Peto explored the nature of transference and in a clinical and everyday context. How did Peto characterise 'transference'? It was the process whereby 'the analyst tried to melt the patient's rigid neurotic patterns in the heat of the transference. Thus the patient go the opportunity to exchange the fixed forms of infantile origin for more adaptable mature ones'. For Peto, 'the analytical situation revived the past, but...it did not give satisfaction in the old ways'. Abstinence was an important part of therapy, and analysis 'ought to be managed in words and only in words'. There were some individuals who could not accept this aspect of analysis:

There were individuals, some of whom belonged to the vaguely defined class of the so-called psychopaths, who could not bear tensions and so were unable to accept the abstinence. They often broke off the analysis when they found out for certain that there was no room for physical gratification in the analytical situation.[33]

The role of the analyst was also at the centre of his discussions. The phenomenon of countertransference was described in considerable detail. It was also the case that the 'more thorough the analyst's own analysis was, the less the strain on him, the less likelihood of his reacting with his own unconscious infantile patterns and the greater the benefit to the patient'. The analyst's mind should be: 'unbiased and unprejudiced from every point of view, devoid of any neurotic conflict; it should operate as a kind of mirror that gave an opportunity for the patient to visualise in the transference situation all his emotional patterns without their being distorted by the analyst's unconscious reactions'. But to do so, to adopt an 'absolutely neutral attitude would be inhuman, insincere and incompatible with the atmosphere of their transference situation'.[34] The aim however was not to be a machine 'without emotions and should not pretend to be one. His attitude should be one of free-floating observation so that he was able to follow the stream of associations as one relatively detached. His unconscious thoughts might run parallel in this way with those of the patient in order to find contact with them at certain emotionally relevant points'.[35] Peto provided a framework and instruction for countertransference in the following terms:

> The analytical attitude meant to notice at once the signs of counter-transference and to handle them by analysing the analyst's reaction. This appropriate handling would enable the analyst to relax, so that the patient's emotions could provoke – after penetrating the analyst's ego – an archaic, similar reaction in his unconscious. The next step was the decisive one for the therapy, to detach from this merging and to grasp it intellectually.

What was demanded of the analyst was a great 'adaptability in the therapist'. It was the responsibility of the therapist to 'register his own unconscious reactions and their conscious derivatives, and move parallel with all these swings of emotions which differed so much in maturity'.[36] It was by no means an easy task to establish these dynamics, as it was a precarious task to undertake transference. In a presentation to the Sydney Institute of Psychoanalysis in 1953, Peto argued 'that there had to stress that the difficulties of building a steady and reliable transference made the cure a precarious task with obsessional neurotics'.[37]

In a paper published in 1955, Peto shows himself to be a Freudian intellectual, widely read and a thoughtful and engaged analyst. In discussing depersonalisation, he draws on the work of Goheim, Lacan,

Klein, Balint and Ferenczi. In issues and problems relating to depersonalisation, they point to difficulties in primary child–mother relations, relating to the missed nipple, anxiety related to this phenomena and oral satisfactions.[38] Peto attempted his own form of psychoanalytic technique. While he recognised the impossibility of reproducing the analytic situation in his articles, Peto engaged theoretically with the practice of analysis and experimented with his own technique:

> I attempted to apply these theoretical considerations while in charge of a psychotherapeutic out-patient department in Budapest where I saw my patients twice a week for fifteen minutes. My aim was the stirring up of unconscious material and breaking through the resistances in order to achieve quick transference phenomena.[39]

His solution in order to achieve this was novel:

> Thus I requested the patients to sit daily in front of a mirror for about fifteen minutes, and to look at themselves without any purpose, just letting themselves go along with their thoughts and emotions. About 50 per cent promptly refused to follow my advice, they found it 'silly', 'strange', laughed with a mixture of embarrassment and aggression. Those who tried had to give it up after one or two attempts'.

Although this had no therapeutic value, it was his way of exploring Lacan's theory about the mirror phase, in relation to his view of the reverse of Lacan's argument about the infantile ego and its integration.

Peto's exploration attempts to examine his idea of 'depersonalisation' – of literally 'falling to pieces'.[40] In drawing on a range of theorists, Peto developed this theory of depersonalisation.

In moving to the USA in 1956, Peto joined a number of immigrant analysts who made the US their home. There, he would continue his analytic work in ways which transcended any connection to the 'nation'.

New York

When Peto arrived in New York, he became a member of the New York Psychoanalytic Institute and the Department of Psychiatry, Albert Einstein College of Medicine of Yeshiva University. At the university, Peto would have found a very supportive and enthusiastic group of colleagues. As Douglas Kirsner has illustrated, there was a strong analytic presence in this department, especially surrounding Milton Rosenbaum, the founder and the Chair of the Psychiatry Department. Rosenbaum envisaged a strong presence of psychoanalysis in the department. There was a close connection indeed with the New York Institute with the department. Kirsner observes that the curricula of the Institute and the Einstein College were identical.[41]

Many of the themes Peto explored in Sydney he also pursued in the New York. His 1959 study explored the ways in which young people from the ages of 15 to 25 discussed their bodies and how this was manifested in the transference relationship. The body image, and symbolism around the body, for Peto became 'signals of the gratifications and frustrations of the child–mother relationship'. There is a development of symbols which 'are closely connected with particular emotions'.[42] These were young people with histories of 'breakdown' in various forms: schizophrenia, anorexia nervosa and anxiety, all of which found expression through a description and analysis of the body. Transference is a key aspect of this dynamic.[43] It was a major theme in his writings and his reflections on transference formed a key aspect of his discussions. In 1960, he wrote of his experience with a patient who had found it difficult to settle down in a trade or a job. Four sessions weekly and six months into his analysis, Peto wrote of his attack with a sense of humour:

> He began with one of his usual attacks on me. His previous attacks were directed against my ignorance of American culture, my stupidity, my incompetence as an analyst ... my inability to point out anything to him he had not known for ten years. The attack in this session was provoked by my necktie. He considered it tasteless and commercial. This in his own words 'worried him no end'. He soon worked himself in his bitterness against my bad taste.[44]

Peto analysed this as 'teasing anger' was the 'only emotion he had allowed himself to develop and this was the only weapon against adults. Other emotions get out of hand, such as sorrow, tenderness or rage.[45] Peto observed that he may have had bad taste, but in fact, it was 'connected with the general pattern of his object relations' and he explored this in relation to his projection of the father–son relationship.[46] In his later publications throughout the 1960s, Peto covered a range of themes and theories in his work such as schizophrenia; dream analysis; anxiety and affect control.

The transference relationship, which Ferenczi did so much to elaborate on and develop remained a particular interest in his work at this time. His paper on 'Body Image and Archaic Thinking' published in 1959 was delivered as a paper to the New York Institute in 1957. The responses by a range of analysts was extremely favourable to his reading of the transference process and in New York, as elsewhere, he found a community of psychoanalytic intellectuals. Phyllis Greenacre discussed Peto's analysis of the transference in the following way:

I am very interested in Dr. Peto's description or limited description of his handing of the transference experience, which I would judge involved a tranquil presence with just enough activity to indicate responsiveness rather than absence; and I would think his finding of even temporary improvement after such sessions may be of great significance...I would think that one of the great gains of Dr. Peto's treatment would be the re-experiencing of the archaic state in an atmosphere of warmth and acceptance leading over directly to verbalisation and other contacts with the analyst...[47]

This particular discussion continued and generated much discussion among members.

One of the most striking aspects of Peto's biography is his writings and intellectual engagement with psychoanalytic ideas. When introducing Peto in 1978 to assembled members of the New York Psychoanalytic Society, Leo Stone described him as a 'distinguished contributor to the psychoanalytic literature'. He described the themes of his papers under the following headings: mother–child relations; on archaic thinking and ego functions; on forerunners of the super-ego; on issues of applied psychoanalysis; other topics such as identical twins, weeping and laughter; war neurosis; delinquency; and body image aspects of depression.[48] Peto's intellectual work was characterised by 'hardy independence and integrity of thought and attitude', which was 'unostentatious, effective, never grossly belligerent'.[49] Stone observed the way in which Peto had enriched the psychoanalytic community in New York.

Several presentations in New York reflect the intellectual and engaged approach he took to his training and work. Even as president of the Society in the 1970s, when he had administrative responsibilities, it was the intellectual work which preoccupied. Hannah Peto recalls how 'he was a man who couldn't stand administration. He wanted to work. He wanted to publish...he liked his patients and he disliked everything which had to be done'.[50]

* * * * *

What characterises these and other writings is an engaged and reflective examination of his patients, as well an intellectual, and not only clinical exploration of the meaning of psychoanalysis. In producing his case studies and analysing them, Peto offers a rare glimpse of the thought processes of an analyst and how he brought theoretical perspectives to bear on the literature.

Peto was an international Freudian intellectual whose work travelled across three countries. In all of these contexts he found a vibrant, engaging community and ready audience for these ideas and arguments, his interest in children being a constant in his psychoanalytic work and ideas. Peto himself ended working within a culture which was not his own. As a European, he worked within an Anglophile culture, but Freudian analysis provided a framework where it appears he could transcend these cultural differences, at the very meeting of the national and the transnational. He analysed his patients within a medical context which was removed from any cultural concerns. Later work by analysts have identified this as a key issue of concern in how psychoanalysts have needed to recognise these cultural shifts and specifications.

The migration of psychoanalysts also raises the issue of the mobility and a cross cultural dimension to psychoanalytic practice. Is it defined by national borders? Certainly, each nation has its own historical development as to how it developed and was sustained by psychoanalytic institutes and cultures. The migratory patterns of analysts have significantly influenced these developments. But beyond the institutions, ideas and practices of psychoanalysis have transcended national borders.

Notes

1. Leo Stone, 'Introduction of Lecturer, Dr. Andrew Peto' Brill Lecture, November 1978; The Abraham A. Brill Library, The New York Psychoanalytic Society, New York.
2. See Joy Damousi, 'Geza Roheim and the Australian Aborigine: Psychoanalytic Anthropology During the Inter-War Years' in Warwick Anderson and Richard Keller (eds) *Unconscious Dominions: Histories of Psychoanalysis, Empire, and Citizenship*, Durham: Duke University Press, 2008.

3. See Mariano Ben Plotkin, *Freud in the Pampas: The Emergence and Development of a Psychoanalytic Culture in Argentina* (Stanford, Stanford University Press, 2001); Joy Damousi, *Freud in the Antipodes: A Cultural History of Psychoanalysis in Australia* (Sydney, University of New South Wales Press, 2005); Alexander Etkind's, *Eros of the Impossible: The History of Psychoanalysis in Russia* (Westview Press, 1997); Nathan Hale, *Freud and the Americans* (New York, Oxford University Press, 1970–1995); Eli Zaretsky, *Secrets of the Soul* (Knopf, New York, 2004). Elisabeth Roudinesco's two volume La bataille the cent ans.
4. Application for Associate Membership/Regular Membership November 17, 1958; Brill Lecture, November 1978, 'Introduction of Lecturer, Dr. Andrew Peto' by Leo Stone; The Abraham A. Brill Library, The New York Psychoanalytic Society and Institute; The Reminiscences of Mrs. Hannah Peto, August 2, 1993, p. 2, Oral History Collection of Columbia University, Columbia University Oral History Research Office.
5. 'AHR Conversation on Transnational History', *American Historical Review*, vol. 112, no.4, December 2006, p.1446.
6. Ibid.
7. Ann Curthoys and Marilyn Lake (eds), *Connected Worlds: History in Transnational Perspective*, Australian National University ePress, Canberra, 2005, pp. 5, 11.
8. Alan Roland, *Cultural Pluralism and Psychoanalysis: The Asian and North American Experience*, Routledge, New York, 1996, p. 21.
9. Ibid., p. 83.
10. Gyorgy Vikar, The Budapest School of Psychoanalysis', In Peter L Rudnytsky *et al.* (eds) *Ferenczi's Turn in Psychoanalysis*, New York University Press, New York, 1996, pp. 65–74.
11. Judith Meszaros, 'The Tragic Success of European Psychoanalysis: The Budapest School', *International Forum of Psychoanalysis*, vol.7, no.4, December 1998, pp. 207–208.
12. Andrew Handler, *A Man for All Connections: Raoul Wallenberg and the Hungarian State Apparatus, 1944–1945*, Praeger, Wesport Connecticuit, 1996, p. 40.
13. See Peto's description. 'Contributions to the Theory of Play,' *British Journal of Medical Psychology*, vol. 29, 1956, p. 100.
14. Stan Gold, 'The Early History', *Meanjin*, vol.41, no.3, September 1982, pp, 342–351.
15. Judith Meszaros, 'The Tragic Success of European Psychoanalysis: The Budapest School', *International Forum of Psychoanalysis*, vol.7, no.4, December 1998, p. 211.
16. Interview with Andrew Peto, with Drs. R. Spiegel and A. Feiner, November 20, 1984, The Abraham A. Brill Library, The New York Psychoanalytic Society, New York, p. 11.
17. Ibid., p. 18.
18. Ibid., p. 15.
19. Ibid., p. 11.
20. Egon F. Kunz, *Blood and Gold: Hungarians in Australia*, F.W. Chesire, Melbourne, 1969, p. 192.
21. See Damousi, *Freud in the Antipodes*.
22. Damousi, *Freud in the Antipodes*, p. 191.

23. *Medical Journal of Australia*, December 22, 1951, vol. 2, p. 862.
24. Ibid.
25. Ibid.
26. Ibid, p. 863.
27. *Medical Journal of Australia*, January 16, 1954, p. 95.
28. Ibid.
29. Ibid.
30. Peto, 'Contributions to the Theory of Play,', p. 104.
31. *Medical Journal of Australia*, July 7, 1951, p. 29.
32. Ibid., p. 28.
33. *Medical Journal of Australia*, July 12, 1952, p. 69.
34. Ibid., p. 69.
35. Ibid., p. 69.
36. Ibid., p. 70.
37. *Medical Journal of Australia*, April 4, 1953, p. 491.
38. Andrew Peto, 'On So-Called 'Depersonalisation', *International Journal of Psychoanalysis*, vol. 39, 1955, pp. 379–385.
39. Peto, 'On So-Called 'Depersonalisation', p. 38.
40. Ibid., p. 38.
41. Douglas Kirsner, *Unfree Associations: Inside Psychoanalytic Institutes*, Process Press, London, 2000, p. 23.
42. Andrew Peto, 'Body Image and Archaic Thinking', *International Journal of Psychoanalysis*, vol. 40, 1959, pp. 223, 230.
43. Ibid, pp. 223–230.
44. Andrew Peto, 'On the Transient Disintegrative Effect of Interpretations', *International Journal of Psychoanalysis*, vol. 41, 1960, p. 414.
45. Ibid., p. 414.
46. Ibid., pp. 414–415.
47. Discussion of Dr. Peto's paper on 'Body Image and Archaic Thinking', November 12, 1957 by Phyllis Greenacre, New York Institute of Psychoanalysis, p. 3.
48. Stone, 'Introduction of Lecturer, Dr. Andrew Peto.'
49. Ibid.
50. The Reminiscences of Mrs. Hannah Peto, August 2, 1993, p. 2, Oral History Collection of Columbia University, Columbia University Oral History Research Office. The author would like to thank the assistance of the Columbia University Oral History Research Office in preparation of this chapter.

6
Psychoanalysis, Transnationalism and National Habitus: A Comparative Approach to the Reception of Psychoanalysis in Argentina and Brazil (1910s–1940s)

Mariano Ben Plotkin

Psychoanalysis is a clear example of a transnational system of thought. Since its creation in late imperial Vienna at the end of the nineteenth century, and particularly after World War II, the center of production and consumption of psychoanalysis has shifted, from continental Europe to the Anglo-Saxon world; and then in the 1960s, to the Latin world (France, but particularly Latin America). In the early 1940s Ernest Jones, then the president of the International Psychoanalytic Association (IPA), told members of the newly created Argentine Psychoanalytic Association that the German language was yielding its place to English as the official language of psychoanalysis. Today Spanish and French are probably the main languages in which psychoanalysis (particularly in its Lacanian version) is produced, discussed and practiced. This displacement from one continent to another and from one language to another had an impact not only on psychoanalysis as a body of thought and as a system of beliefs, but also on the different cultural spaces in which it took roots and developed.[1]

This chapter is an attempt at articulating both the transnational and the national dimensions of the development of psychoanalysis. I would like to propose that the transnational aspect of a system of thought (psychoanalysis in this case) cannot be properly understood without taking into consideration, in a comparative fashion, national patterns of reception and circulation. This chapter compares the early reception of psychoanalysis in two Latin American countries in whose urban culture it has had a deep impact: Argentina and Brazil. The interest is not placed on the institutional history of psychoanalysis, but on its multiple levels of reception and interpretation that developed independently from the establishment of an orthodoxy. I am thus interested here in psychoanalysis defined in broad terms, as a cultural artifact and not just as a formalized psychological theory or a therapeutic technique.[2] I will focus here only on three of the multiple areas of reception of psychoanalysis which I consider relevant: the medical establishment, the avant-garde literary circles and the social sciences. What is usually considered as "popular culture" constitutes another important space for the reception and diffusion of psychoanalysis, the analysis of which raises interesting methodological issues, which I have dealt with elsewhere for the case of Argentina.[3] Although casual references will be made to other regions of both countries, I will concentrate mostly on Buenos Aires, the undisputed cultural and economic center of Argentina, and on São Paulo and Rio de Janeiro, the two poles of cultural modernization of Brazil. I will show how specific patterns of reception of psychoanalysis in these two countries are linked to aspects of the social imaginary associated to the construction of national identity; in particular to what Norbert Elias has defined as "national habitus," that is to say the way in which "the fortunes of a nation over the centuries become sedimented into the habitus of its individual members."[4] Elias's concept of habitus articulates the macro and the micro levels of analysis. I would like to propose that in Brazil and Argentina, a habitus that gradually formed during the nineteenth century shaped the early reception of psychoanalysis, giving particular nuances to the transnational nature of the Freudian doctrine.

Argentina and Brazil: two countries so close and so far away

Since the nineteenth century Argentine and Brazilian intellectuals have been looking at each other's countries trying to find a model for what their own country should or should not be like. After achieving independence from Spain in 1816, Argentina entered into a long period of civil wars and dictatorships that formally ended in 1853 when a constitution was finally drafted after the fall of dictator Juan Manuel de Rosas. From then on, the country entered into a long era of economic prosperity and political stability that lasted, with ups and downs, until 1930 when a philo-fascist coup d'etat and the consequences of the economic crisis of 1929 put an end to it. Since the second half of the nineteenth century, European capital and immigrants had flowed into the country changing the ethnic composition of the population.[5] By 1914 over one half of the adult male population of the city of Buenos Aires (which housed almost one third of the entire country's population) was foreign-born, meaning mostly European. In the meantime, the city bloomed, becoming in the eyes of locals and foreign visitors alike a truly European and cosmopolitan enclave in South America. In the words of Richard Morse, by the 1910s Buenos Aires seemed to have crossed the threshold of Western Modernization.[6] Ethnically and culturally, *porteños* (the habitants of the port-city of Buenos Aires) considered themselves to be closer to Europe than to any other Latin American country.

Nonetheless, from the nineteenth century political polarization has become a crucial component of Argentine's habitus. The "us/them" dichotomy, usually conceptualized in terms of "civilization (i.e. European) vs. barbarism (i.e. native or mestizo)" had provided an interpretive grid for Argentine intellectuals since the times of Domingo Sarmiento's seminal essay *Facundo* (1845).[7] The arrival of massive immigration, however, forced local elites to reconsider the issue of national identity from a different perspective. The fast transition of Buenos Aires from a "gran aldea" (great village) to a modern city was traumatic. By the turn of the century social and intellectual elites started to perceive the modern city as a space of corruption and of degeneracy. Rising levels of crime, deteriorating material conditions, the emergence of a combative urban proletariat and, particularly, the feeling that immigration degraded not only the "true" national culture, but also eroded the social position until then monopolized by the native elite – also of European origin, but established in the country for a longer period of time – generated a wave of active nativism among intellectuals.[8] Nonetheless, the elites were well aware of the country's need for workers in its blooming economy and thus nationalism did not lead to the imposition of quotas or other kinds of legal limitations to immigration. Rather, the influence of neo-lamarckism among the local elites promoted social policies, particularly an emphasis on the massive expansion of public education, which was conceived as a major tool for the integration of immigrants. In the 1840s, intellectuals such as Sarmiento (President between 1868 and 1874) and Juan Bautista Alberdi had promoted European immigration as an instrument for civilizing and educating the "wild Pampas," but by 1910 President of the National Council of Education and psychiatrist José María Ramos Mejía used the Argentine system of public education to "civilize" and "Argentinize" the "primitive" Europeans (mostly Italians and Spaniards) that were arriving in masses in the port of Buenos Aires. The "civilizers" of yesterday had become the "barbarians of today." Later, particularly after the emergence of mass politics in the 1910s, but especially since the peronist decade (1945–1955) this dichotomous view of society was reformulated in political terms.

Brazil's history is very different. It was the only country in the Americas to keep a monarchy after achieving its relatively non-traumatic independence from Portugal in 1822.[9] The monarchy secured for Brazil a level of political stability that was unknown in the emergent republics of the fragmented former Spanish colonies. The survival of the monarchy was closely associated to slavery, which, for centuries, had been the backbone of the workforce of the Brazilian plantation economy. Although nineteenth-century Brazilian social and intellectual elites held, at least formally, a liberal credo, they were aware that republican practices and discourses could be hardly compatible with the presence of a large population of black slaves.[10] Monarchy outlived the abolition of slavery in 1888 by only one year: in 1889 Emperor Pedro II was overthrown by a military revolution that established a republic more in tune with the rest of the continent. This oligarchic republic (know as the "old republic") lasted until 1930 when a revolution led by Getulio Vargas ended it, thus beginning a new era of centralized authoritarianism.

Since the nineteenth century, modernization had been for Brazilian intellectual elites a synonym of "whitening" the country.[11] Doctrines of "scientific" racism originating in the ideas of Count of Gobineau (who served as Ambassador of Napoleon III to the Brazilian court), Louis Agassiz and a particular reading of Social Darwinism were combined with the idea that a gradual but fast "whitening" of the population was possible through European immigration and competition for survival, since the "superior" white race would prevail. Thus, most educated Brazilians considered blacks and mulattoes as degrading elements in society.

Brazilian intellectuals looked at Argentina as a successful model to follow. Although a large number of foreign immigrants had gone to Brazil since the late nineteenth century, the country was never as successful as its neighbor in attracting the coveted white Europeans. Thus, in Brazil the formulation of national identity was closely related to the possibilities and limits of integrating the different ethnic groups that composed its society, particularly aborigines and blacks. However, while the former (less visible in the cities) had been idealized by a generation of nineteenth century romantic poets and writers, the latter were despised as an inferior group. In general, the presence of a sizable black and native population was perceived by members of the Brazilian elite as a hindrance to the country's progress, and as a problem that had to be urgently addressed.[12] However, high levels of miscegenation had existed in Brazil since colonial times, giving rise to a large colored population and to the myth of a Brazilian racial democracy.[13] Unlike the US, Brazil developed as a multiracial society in which possibilities of social mobility were closely tied to "whiteness." Moreover, "whiteness" could be achieved by social prestige. It was possible for Brazilian mulattoes and even for blacks to rise to the top of society.[14] While Argentine elites were relatively successful in constructing an image of their country as a European cosmopolitan enclave, in Brazil the "exotic" elements that decades later would fascinate European avant-garde artists were part of everyday life. Brazil became an icon of exoticism in Latin America.

The first decades of the twentieth century were a period of great disruption in the country. Former slaves emancipated in 1888 were left without education or any kind of social safety net. Many of them emigrated to the cities giving origin to a colored urban underclass that fed the incipient but fast-growing industrialization of Brazil. As a result Brazil, a traditionally rural country, went through a process of fast urbanization. The population of the city of São Paulo almost doubled between 1893 and 1900 while the city modernized rapidly. European immigration that started pouring into the country after abolition also greatly contributed to this urban growth.

Thus, in Brazil the formulation of national identity was closely related to issues concerning race relations and the need of dealing with different ethnic groups that composed its society, particularly aborigines and blacks. For the Argentine elites, on the other hand, the problem of national identity was formulated quite differently. Argentine elites were obsessed with the problem of integrating the immigrants and looking for the "true roots" of a country whose culture and ethnic composition was changing fast. This search for a national identity was superimposed with political polarization and a dichotomous view of society that regained its strength in times of crisis.

Medical reception of psychoanalysis

The first reception of psychoanalysis in medical circles took place in both countries at the beginning of the twentieth century. The positivist paradigm had been extremely influential since the last decades of the nineteenth century – as in the rest of Latin America.[15] In psychiatry this paradigm was related to what Nathan Hale has called "the somatic style,"[16] that is to say, the idea that mental diseases were determined by the morphology of the brain. The beginning of the twentieth century, however, saw positivism begin to decline, a decline that accelerated after World War I. The reception of psychoanalysis both in Argentina and Brazil, has to be understood in the context of the crisis of the positivistic paradigm, crisis that left a space open for alternative interpretive models. Although there was a positivistic interpretation of psychoanalysis that emphasized its biological and "deterministic" dimensions, it was generally understood as an option to previous purely somatic theories. Degeneracy theory, for instance, was a major current in Argentine and Brazilian psychiatry until well into the twentieth century.[17] This theory, formulated in nineteenth century France was based on the idea that mental and physical diseases were passed from generation to generation, each time in a more destructive dose. In Argentina this theory became associated with immigration. Members of the Argentine social and cultural elite argued that uncontrolled immigration would deteriorate not only the national culture but also the national "racial stock." For a variety of reasons, immigrants constituted a large proportion of patients in mental hospitals and the "crazy immigrant" became an iconic figure in literature and more generally in the social imaginary.[18]

In Brazil degeneracy was mainly linked to "blackness." According to Raymundo Nina Rodrigues the practice of African religions and rituals were evidence of the "extreme neuropathic or hysterical and profoundly superstitious personality of the Negro." Others thought that Spiritism practiced by blacks and mulattoes "could induce madness in any participant individual, with or without a predisposition."[19] Through their religious rituals and also, and perhaps more importantly, through their exaggerated sexuality that led to perverse practices, the blacks and the mix-raced population introduced the germ of degeneracy into "Brazilian" society. Race and sexuality, therefore, were closely linked in Brazilian social and medical thought.

These intellectual contexts shaped the early reception of psychoanalysis in each country. In Argentina, a country where French cultural influence was paramount, psychoanalysis was originally read in French (meaning through French sources or, more commonly, through French commentators) and interpreted through the lens of French theory.[20] Germán Greve, probably the first doctor who mentioned Freud in Argentina, in a paper delivered at the International American Congress of Medicine and Hygiene in Buenos Aires in 1910 ended his presentation by saying: "and let us compare Freud's opinion about the primary etiology of neuroses with the one [Pierre] Janet has expressed on the same question, because we need to note the agreements existing between the two in order to reconcile [Freud's theory] with such a distinguished opinion [as Janet's] ... "[21] If Greve tried to legitimize the introduction of Freud's theories by showing that his points of view were compatible to those of Pierre Janet, others would use the French psychologist to criticize psychoanalysis. José Ingenieros, for instance, a prominent psychiatrist, psychologist and intellectual active in the first three decades of the twentieth century, introduced a short discussion on psychoanalysis in the 1919 edition of his best selling *Histeria y sugestión*. There, Ingenieros criticized Freud's theories using the very same arguments presented by Janet in a polemic text of 1913: Freud overemphasized the role of sexuality in the etiology of neuroses, thus taking the discussion on mental disorders out of the realm of scientific medicine.[22] Similarly, Dr. Alejandro Raitzin, a respected forensic psychiatrist published an article on madness and dream in 1919, where he offered an extensive critique of Freud's theories. At the end of the article, however, Raitzin recognized that his knowledge of psychoanalysis was limited to his reading of Emmanuel Régis and Angelo Hesnard's *Psychanalyse des névroses et des psychoses*, published in France in 1914.[23] Even doctors who approached psychoanalysis with sympathetic eyes considered that it was useful only in what it had in common with French psychiatric thought. Like the French doctors, they rejected Freud's "pansexualism," and had less problems in accepting Freud's "technique" than in tolerating his "theory." Psychoanalysis was welcomed by some Argentine doctors in the first decades of the century as long as it was de-sexualized.

Prominent Brazilian doctors seemed to have been more receptive to psychoanalysis than their Argentine colleagues and, unlike them, focused on sexuality as psychoanalysis came to replace, to some extent, other theories that also placed sexuality at the center of their interest. Therefore, the Brazilians emphasized what the Argentines repressed. While for the Argentines the word "pansexualism" had – following the French school – a negative connotation, in Brazil it was the opposite. In 1920 Franco da Rocha published his *O pansexualismo na Doutrina de Freud*,[24] where he had a sympathetic view of the basic concepts of psychoanalysis. Ten years later, a second edition was published under the title *La Doutrina de Freud*. This time da Rocha decided to eliminate the word "pansexualism" after learning the negative connotation that such a word had in Europe. For Rocha, far from signifying a criticism to Freud, "pansexualism" defined psychoanalysis.[25]

It is noteworthy that while the Argentine doctors who became interested in psychoanalysis in the 1920s and 1930s were in general younger members of the profession who occupied a relatively marginal position in the field, in Brazil some of the most prestigious doctors, who were more familiar than their Argentine counterpart with the German language, were attracted to Freud's theories. As early as in 1899 (that is to say, even before the publication of *Interpretation of Dreams*), Dr. Juliano Moreira began discussing Freud's writings in his classes. Although he did not publish specific works on psychoanalysis, his interest in it had a strong influence among some of his students. During Moreira's term as director of the "Hospital Nacional de Alienados" (Brazil's national mental hospital), between 1903 and 1930, he established a "psychoanalytic ward."[26] In the 1910s medical students started writing dissertations on psychoanalysis while courses discussing it became available at the Rio de Janeiro medical school in 1918. At the same time other prestigious doctors in São Paulo, such as Franco da Rocha, also became interested in psychoanalysis.[27] Both Moreira and da Rocha became members of the short lived Brazilian Psychoanalytic Association created in São Paulo by Dr. Durval Marcondes, a student of Rocha, in 1927. This association received provisional recognition from the IPA and published a single issue of a journal. However, it disappeared one year after it was created. Its members included prestigious physicians but also artists and intellectuals not linked to the medical profession and not interested in pursuing a professional career in psychoanalysis.

In Argentina, political polarization had an impact in the development of psychoanalysis. Psychoanalysis was read and interpreted in political terms. When psychoanalysis became part of the Argentine culture it also became an object of political appropriation. By the time Spanish doctor Ángel Garma and a group of followers created the Argentine Psychoanalytic Association in 1942, most doctors who had been interested in Freud had already rejected psychoanalysis mostly for political reasons. Left-wing doctors who had been sympathizers or members of the Communist Party, such as Gregorio Bermann – who had published in the 1930s in a journal called *Psicoterapia* and who had lectured extensively on psychoanalysis since the 1920s – or Jorge Thénon – who had written his dissertation on a topic related to psychoanalysis and had sent it to Freud for his approval – had seen in psychoanalysis a tool for social reform and for the renovation of psychiatry. However, by the early 1940s they had abandoned their earlier interest in Freud's ideas, following the dictates of the Soviet Communist Party or, more importantly, the tendencies followed by French communists who condemned psychoanalysis as an idealistic-bourgeois doctrine. Similarly, right wingers attracted to more or less fascist ideologies, such as forensic doctor Juan Ramón Beltrán, who had also shown an interest in the Freudian system but saw it as a tool for social control, rejected psychoanalysis as well because he, and others like him, perceived it now as a subversive or Jewish science. As a result, and unlike in Brazil, none of the doctors who in the previous decades had been associated with the practice or diffusion of psychoanalysis joined the new association. Thus, whereas the Brazilian psychoanalytic movement built for itself a genealogy that traced its origins to the early reception of psychoanalysis, nothing like this happened in Argentina where psychoanalysts quickly became a self-referential group, at least until the 1960s.

The combination of both foreign and domestic historical events such as the Spanish Civil War, the raise of fascism, World War II, the de coup d'etat of 1930 and, above all, the emergence of peronism in the 1940s deepened the already existing political polarization during the decades of 1930s and 1940s. This schism affected all dimensions of the country's social life, including the scientific and medical fields. Until the mid-1930s leftist Bermann and rightist Beltrán could still share institutional and social spaces for the discussion of psychiatric and psychoanalytic issues – Beltrán, for instance, was a member of the editorial committee of Bermann's journal *Psicoterapia* – later in the decade this kind of "pacific coexistence" became impossible. Most fields of social interaction became "cannibalized" by politics. Moreover, after the rise of Perón to power in 1945, psychoanalysis was perceived as part of the liberal cultural system that opposed peronism. Although most psychoanalysts refrained from open political activity during that time, it became clear both for the government and for the psychoanalysts themselves that psychoanalysis did not fit into the peronist cultural programs. Unlike their Brazilian counterparts, Argentine doctors who promoted the use of psychoanalysis were routinely fired from public hospitals. Consequently, until the 1960s, Argentine psychoanalysts became a self-contained community of medical doctors who ended up cutting their ties with larger medical and intellectual circles.[28]

Things developed quite differently in Brazil where psychoanalysis received a much more homogeneous reception in medical circles. Psychoanalysis allowed doctors and intellectuals who were unhappy with the prevalent racially deterministic view to introduce some nuances into it. As Jane Russo points out, while racial theories condemned the country to failure and backwardness, psychoanalysis provided a way out.[29] If the uncontrolled sexuality attributed to the blacks had been seen as evidence of primitivism, psychoanalysis provided new tools to approach the problem. First, the idea of sublimation, interpreted through the neo-lamarckian lense, introduced the possibility of "educability." Thus, the excessive sensuality of primitive races needed not be an impossibility for progress. Second, the very concept of "sexual excess" and primitivism came to be seen under a different light because in themselves they were neither good nor bad as psychoanalysis had shown. Sexuality, under the new perspective, was responsible both for the worse and the best of what humanity could do, while it was recognized that everybody had a "primitive" dimension in his/her personality. Some thinkers found in sexuality the origins of Brazilian identity.

In the decades following World War I, and particularly after the Revolution of 1930, some Brazilian intellectuals and doctors attempted to construct a notion of national identity that would be more inclusive than the prevalent racist conceptions. Social thinkers such as Paulo Prado, Sergio Buarque de Holanda and Gilberto Freyre, among others, wrote influential essays with the aim of analyzing (as the title of historian Buarque de Holanda's book of 1936 suggests), the "roots of Brazil."[30] Particularly important was the path-breaking socio historical analysis that Gilberto Freyre, a leading sociologist, wrote in the early 1930s: *Casa grande e senzala* where he dismissed the "racialist" interpretations of Brazilian historical evolution. According to Freyre Brazil's strength was based on the fact that it was a "mestizo" country constructed upon the illicit sexual relations between the (male) white master and the (female) black slave within the context of traditional patriarchal family. Although Freyre did not directly discuss Freud (he did discuss Oskar Pfister's ideas, however), his emphasis on sexuality as a component defining national identity was compatible with a discussion of this and related issues in psychoanalytic terms.[31]

In the Brazilian doctors' point of view all Brazilians, independently of their race had a "primitive ego," identified with the id, that needed to be disciplined, civilized and transformed.[32] It shouldn't surprise, therefore, that the early reception of psychoanalysis in Brazil was closely related to education and to a neo-lamarckian reading of Freud's ideas. The lamarckian emphasis on evolution through inheritance of acquired characteristics, which could be induced from outside, left room for optimism since Brazilian "primitivism" could be "civilized" through education and other state sponsored measures. Artur Ramos, from Bahía, who was active in the fields of medicine, education and anthropology, and a student of black culture in Brazil, created, following Jung, and inspired on Lucien Lévy-Bruhl's work on "primitive mentality," the concept of "folkloric unconscious" which would explain the persistence of primitive mental elements in contemporary man. This folkloric unconscious became visible under certain conditions, such as dreams, mental disorders and artistic creations. Ramos refuted the theses concerning the racial inferiority of the black population. Instead, he proposed an ethnopsychoanalytic diagnosis of the Brazilian people, imprisoned, according to him, in a "mental atavism" originating in the influence of Afro-Brazilian culture evident in some fetichist rituals. The condition of the Brazilian black was, according to Ramos, "the consequence of magic, pre-logical thought, which is independent from the anthropological-racial question, because it can emerge, under different conditions, in any other ethnic group...This idea of 'primitivism,' of 'archaic' is purely psychological and has nothing to do with the issue of racial inferiority."[33]

Some prestigious doctors, most notoriously Júlio Pires Porto-Carrero (born in Pernambuco but active in Rio de Janeiro since the late 1920s),[34] saw in psychoanalysis a tool for the regeneration of the population, particularly through state sponsored sexual education and eugenic measures, including a liberal policy toward abortion.[35] Since Brazilian homes, affected by the vestiges of primitivism carried by local families, could not secure social discipline and civilization, it was up to the state to carry out this task through (among other things) the sublimation of children's sexual impulses by imparting sexual education. As Porto-Carrero pointed out in 1927: "Given the influence of sexuality in the formation and operation of the infantile psyche, it is not fair that education ignore the sexual side of life and repel sexual manifestations and knowledge simply as immoral. It is urgent to introduce sexual education."[36] The price of ignoring psychoanalysis in education would be crime and perversion. This is why Porto-Carrero insisted on educating parents in psychoanalysis. Once society is rebuilt on the foundations laid by psychoanalysis, crime would cease to exist.

Porto-Carrero combined psychoanalysis and criminology. While earlier criminologists like Nina Rodrigues linked criminality to race, arguing that blacks and mulattoes could not be considered fully responsible for their acts and therefore could not be held legally competent to stand trial in a civilized court of law, Porto Carrero displaced the question of criminal responsibility from race to the realm of the unconscious.[37] For him no criminal could be held responsible for his or her acts, since criminal actions were the consequence of problems located in the unconscious. Moreover, the criminal was, according to Porto-Carrero, a social emergent: "the criminal embodies the guilt and sadistic anxieties of the whole society."[38] Porto-Carrero proposed a general reform of the penal code that would eliminate punishment for criminals, since according to him, punishment was an act of revenge inflicted by society originating in a feeling of collective guilt rooted in the Oedipus Complex.[39]

Several Brazilian doctors who engaged with psychoanalysis in the 1920s and 1930s were active in their states' educational systems. Dr. Ulisses Pernambucano, for instance, was director of the Escola Normal de Pernambuco, Dr. Durval Marcondes from São Paulo created and led the Mental Hygiene section of the Paulista school system in the 1930s, while Dr. Artur Ramos was director (1934–1939) of the "Secão de Ortofrenia e Higiene Mental" of the Institute of Educational Research, set up as part of a broader program for educational research promoted by the Getulio Vargas government after 1930. All of them had a "pedagogical" idea of psychoanalysis. It is hardly surprising, therefore, that the Freudian category of "sublimation" occupied a key place in the Brazilian doctors' reception of psychoanalysis, since sublimation meant the re-direction of sexual drives to other ends. Education was in the eyes of these doctors, a state sponsored tool for sexual education and sublimation.

Unlike Perón 15 years later, Getulio Vargas attracted intellectuals regardless of their ideological colors.[40] Patterns of collaboration between intellectuals, most of them belonging to elite families and the state had existed in Brazil since the times of the monarchy.[41] Vargas carried out a program of social and cultural modernization that was much more intellectually (not socially) inclusive than Perón's. This program was compatible to the image that many Brazilian intellectuals had constructed for themselves as the providers of guidance to an incapable and uneducated people. The Vargas regime created a strong, centralized state willing to work with intellectuals on a broad program of political and cultural reforms. Brazilian political and intellectual elites shared much more of their social and cultural background than their Argentine counterparts. Thus, it was easier for Brazilian doctors and educators to introduce psychoanalysis in official educational agencies and hospitals than for their Argentine colleagues. Moreover, psychoanalysis could be appropriated by Brazilian intellectuals and doctors to make it fit into accepted ideas of social reform. In Argentina, on the other hand, there was a historical pattern of mutual distrust between intellectuals and the state that deepened dramatically during the Perón regime. Moreover, during the Perón government psychoanalysis was considered part of the liberal-anti peronist culture. Psychoanalysts, together with avant-garde artists, and social scientists found a place in Vargas's modernizing programs that their Argentine colleagues could not find in politically polarized Argentina.

Reception of psychoanalysis in non-medical circles

Psychoanalysis and avant-garde art

Medical doctors were not the only ones in showing an interest in psychoanalysis. Since the 1920s there was a growing perception that psychoanalysis was not only an innovation in psychology and psychiatry but also an essential component of cultural modernity. In the post-World War I period, Latin American intellectuals started questioning previous certainties. The barbarism unchained in Europe repositioned the whole discussion about civilization and the supposedly primitive nature of Latin American societies. The "primitive" and "savage" aspects of Brazilian culture started to be seen in a new light. This revisiting of old themes had a deep impact on the arts and, more generally, on the Brazilian intellectual field.

In February 1922, during the celebration of the centenary of Brazilian independence, the "Modern Art Week" took place in São Paulo. The eight days of public exhibit followed by three days of festivals (lectures, readings and concerts) were calculated to scandalize the bourgeois public of the city.[42] The Modern Art Week was the crystallization of a movement that had started taking shape in the previous decade around writers such as Mario de Andrade, Oswald de Andrade (no relation to the former), and painters such as Lithuanian born Lasar Segall, Anita Malfatti and Tarsila do Amaral, among others. "Modernism" in Brazil was the equivalent to "avant-gardism" elsewhere and permeated into all forms of the arts.[43]

Although influenced by Dada, Fauvism and Futurism, Brazilian Modernism did not mimic any of those movements. The several Modernista "manifestos" emphasized the idea that Europe had profited from Brazil for a long time and now it was high time for Brazilian culture to "cannibalize" Europe. Cannibalizing meant taking whatever was useful from Europe, and to creatively digest it into something new (and to excrete whatever could not be put to use) or, as Oswald de Andrade put it "Tupy or not Tupy, that is the question."[44] Modernism had its most dynamic moment during the 1920s,[45] when the movement diversified taking different forms in various regions of the country.

Many modernists made heavy use of psychoanalytic concepts and ideas but did so from a point of view that was very different from the one held by the psychiatrists. For the latter, psychoanalysis was a tool for understanding and taming the wild components of Brazilian culture or, in the best of cases, a possible way out from biologic determinism. For the artists, psychoanalysis provided (together with some aspects of Marxism and a reading of Nietszche, among other elements) the foundation for an aesthetic that exalted precisely those exotic and wild elements of Brazilian culture, exactly those that the doctors wanted to repress.[46] The concept of the "primitive" was used by Oswald de Andrade to exalt the matriarchal sexual organization of natives before the colonial patriarchal order was established. Once again, de Andrade emphasized sexuality as a key element in Brazilian national identity. The native, the "mestizo," and the black were re-interpreted as the creative forces of Brazilian culture and society. During his trips to Paris Oswald found that the "primitive" cultures that excited European avant-garde artists were part of everyday life in tropical Brazil. While the European avant-garde movements exalted primitivism as a rupture with national traditions, in Brazil (and to some extent in Argentina too), the recovery of "the primitive" established links with the national past. Latin American avant-gardism thus established a relationship with national traditions and history that was different from those established by their European counterparts.[47]

For the Modernistas, Freud and psychoanalysis provided a diagnosis of social problems but not necessarily a cure. In the "Manifesto Antropófago" of 1928 de Andrade mentioned Freud three times: he was characterized as the man who solved the enigma of women, as the man who identified the evils that antoropofagia came to end, and as the man who understood the oppressive nature of the social reality.[48] Moreover, the idea of the taboo turned into a totem through antropofagia (thus reversing Freud's sequence of a totem turning into a taboo) pervades the whole "Manifesto," which was dated on the 374th year of the "deglutition of the Bishop Sardinha" (a Catholic bishop killed and eaten by the Caeté Indians in 1556).[49] Mario de Andrade's *Macunaíma* (1928), considered by most critics as the best example of the anthropophagic movement, tells the dream-like trip of a tapanhuma Indian from the forest to the modern city of São Paulo, in search of a lost talisman. Freud is also explicitly mentioned in this work which abounds in slips, references to the child present in the adult, and other psychoanalytically inspired elements.

The modernists exalted as the most creative component of Brazilian culture what the psychiatrists sought to tame. Thus, as Cristiana Facchinetti claims, psychoanalysis attracted the interest of modernists and their friends in two opposite ways: as an instrument for criticism of civilization shaped in European molds, and as a device to valorize peripheral cultures and their "primitive" components.[50] "I write without thinking all that my unconscious yells at me" wrote Mario de Andrade at the beginning of his "Préfacio Interessantíssimo" to *Paulicéia Desvairada* of 1920, a long poem about the urban metropolis of São Paulo.[51] In Mario de Andrade's view artistic creation originated in the liberation of the unconscious from archaic structures. Art was a manifestation of the collective unconscious. Sublimation, for instance, a concept that was so central for psychiatrists like Porto-Carrero, was rejected by the Modernistas who proposed instead, a "swallowing up" of the enemy."[52] In the Modernistas' appropriation, however, Freud's ideas were also subjected to "anthropofagia." Modernistas took what they could use of psychoanalysis, combined it with other theoretical frameworks, most notably Ribot's psychology, and re-intepreted it to make it fit into their aesthetic and political agenda. Unlike Surrealists, the Modernistas approached psychoanalysis from a non-theoretical perspective; for them psychoanalysis was an aesthetic and ideological tool. Moreover, in contrast to their French counterparts, Modernistas were not interested in issues related to mental health. While the surrealists exalted the hysteric woman to the role of a muse, Modernistas sought inspiration in the primitive components of Brazilian culture.

In spite of the apparent difference in the uses of psychoanalysis made by artists and doctors, these patterns of reception had something important in common. In both cases the problem addressed was the same: how to deal with the "primitive dimension" of Brazilian culture and society. This problem was central in the constitution of Brazilian national habitus. Given Modernistas' interest in psychoanalysis, it is not surprising that when in 1927 Dr. Durval Marcondes established his Brazilian Psychoanalytic Society, more than a few of the artists who had participated in the Modernist Week joined it immediately.[53]

Argentine avant-garde movements were less receptive to Freud and psychoanalysis than their Brazilian counterparts. In fact, psychoanalysis had no noticeable influence in the artistic avant-garde of that country during the 1920s and 1930s. The sector of writers that could be characterized as "modernizer," close to the politically and socially committed group known as the "Boedo group", was more receptive to psychoanalysis than the aesthetically more advanced group of writers known as the "Florida group."[54] Argentine avant-garde writers, even those most eager to shake up the establishment, such as the groups associated with the journal *Martín Fierro* in the 1920s, were much more moderate in their goals than their counterparts elsewhere. Unlike Brazilian Modernistas and others, their rupturist project was defined purely on aesthetic terms. *Martín Fierro*, for instance, made clear from its first issue its complete detachment from political and social questions.[55] Although also written with the purpose of *épater le bourgeois*, *Martín Fierro's* manifesto is much less ideologically (and also aesthetically) radical than either the Pau-Brazil or the Anthropophagic manifesto. While the Brazilians wanted to reformulate language and create a new aesthetic based on primitivism and exoticism with the purpose of introducing deep changes in culture and society, *Martín Fierro* claimed that it represented a break with "traditional literature," but it recognized at the same time that this attitude was not incompatible with having – "as in the best families, an album of family pictures [of ancestors] to laugh about, but also to revere."[56]

As Beatriz Sarlo has pointed out, the people who participated in the *Martín Fierro* project and its successors wanted to free literature from its social and ideological foundation – they made a point in claiming the autonomy of literature from politics – and in that sense they were at the opposite extreme of such movements as Surrealism, Futurism and even Brazilian Modernism. This project, with its strong antipsychological component, made no use of psychoanalysis. Even Oliverio Girondo's erotically charged poems show no influence of psychoanalytic thought. As literary critic Jorge Schwartz points out, in Girondo poems, sexuality becomes a public act deprived of any psychological content. Eros is taken away from the realm of passion and associated instead to a "technological" and parodic interpretation of sexuality.[57] While Oswald de Andrade individualized the erotic experience, charging it with sensuality, Girondo used a reified notion of sexuality as a parody of established social mores.[58]

Argentine avant-gardes formulated the question of national identity in a very different way than their Brazilian colleagues. There was nothing exotic to celebrate in "civilized" Buenos Aires. What Argentine intellectuals did exalt was their European-style cosmopolitism in a moment of unparalleled economic prosperity and democracy. When Argentine avant-garde writers such as Ricardo Güiraldes or Jorge Luis Borges paid tribute to the primitive creole roots of Argentine culture, they were forced to look at, or even to construct, a past. By the time of their writings, the "gauchos" and the "compadres" that populated their fiction had long disappeared from the Argentine pampas and the city. These writers were reacting against the fast and uncontrolled social change by constructing a nostalgic tradition that negated those changes. Unlike Brazil, in Argentina the main problem associated with the building of a national identity was constituted by the waves of immigrants who distorted the language and customs.[59]

Some elements of psychoanalysis, however, were incorporated in the literature of the Boedo group. As Roberto Mariani, a member of that group, pointed out: "realism in literature has left Zola behind and has detached itself from uncomfortable companies (from sociology, principally....); at the same time it developed vigorously with new or renovated contributions, such as the unconscious."[60] Roberto Arlt, is a case in point. He was a self-taught writer, child of Austrian immigrants who, unlike Borges, Oliverio Girondo, Ricardo Güiraldes and the people associated with *Martín Fierro, Proa* and other rupturist journals, did not posses a solid educative or social capital.

Arlt did incorporate some elements of psychoanalysis in his fiction, most notoriously in his dostoyevskian novel *Los 7 locos* (1929) and its sequel *Los lanzallamas* (1931) where episodes in the childhood of the protagonist were used to explain his sadomasochist tendencies in adulthood and references to "subconscious" fantasies are conspicuously present in the texts. Arlt's use of psychoanalytic concepts, however, should be seen in the context of his appropriation of what Beatriz Sarlo has characterized as the "knowledge of the poor." He incorporated popular notions associated with psychoanalysis in the same way in which he introduced popular technical knowledge: as an alternative, not sanctioned by the establishment, form of knowledge aimed at and appropriated by those lower-middle class sectors who benefited from the expansion of literacy but who were, nonetheless, excluded from the realm of accepted "high" culture. This knowledge circulated through non-academic and informal channels such as newspapers and popular magazines, and constituted the basis for a "technical imagination" in which fiction and reality sometimes overlapped.[61] The possession of this technical and semi-scientific knowledge provided the opportunity, according to Sarlo, for a reorganization of a social hierarchy of knowledge. In the 1920s and 1930s popularized versions of psychoanalysis circulated like those technical, semi-scientific forms of knowledge to which Sarlo refers. Psychoanalysis was presented in popular publications at the same time as cutting edge, almost miraculous medical technique, and as a method legitimized by "science" of approaching old passions such as the interpretation of dreams or the mysteries of the unconscious and sexuality.[62] For Arlt and others, therefore, psychoanalysis was less the source of a new aesthetic or a literary instrument than a component of those informally acquired forms of knowledge that, together with popular technical information, are present in his texts.[63]

Psychoanalysis and the social sciences

The development of the social sciences, which was associated in both countries with local university and intellectual traditions, was another realm where early interest in psychoanalysis was seen. Argentine public universities have a long history dating back to the seventeenth century when the Universidad de Córdoba was created by the colonial government. The University of Buenos Aires, the largest and most prestigious one, was founded in the 1820s, after the independence. Other universities were created later in the interior of the country, the most notable being the Universidad de la Plata. Private universities were not allowed to operate until the late 1950s. Moreover, Argentina was the country of origin for the movement of "Reforma Universitaria", which began in Córdoba in 1918, and which had continental projections. This movement opened the doors of higher learning to the middle and lower-middle classes thus increasing considerably the number of students. It also introduced politics into the universities by establishing the system of co-government in which students, graduates and faculty had each made up one third of the representatives in the governing bodies of the universities.

The situation was different in Brazil where until the decade of 1920 there were no universities. The children of the imperial nobility or of the republican bourgeoisie received their professional training at independent, although mostly public professional schools where they could study engineering, medicine or law. The universities were created as part of the modernizing projects that Vargas implemented at the federal level and the state of São Paulo and others put in practice at the local. However, the most modern and prestigious universities were created in São Paulo after the failed constitutional revolution of 1932 which was organized by sectors of the paulista bourgeoisie against Vargas's policies of centralization. Although the federal troops defeated the revolution the victory of the central government was ambiguous. Vargas knew that São Paulo would continue to be the engine of Brazilian economy and culture and therefore some kind of agreement had to be reached with the rebels.

The failure of the revolution, on the other hand, forced the paulistas to revisit the mechanisms of reproduction of elites and, particularly, to device new institutions for the education of a local intellectual and technical elite. This was the main reason behind the creation of the University of Sãn Paulo (USP) and also of the Escola Livre de Sociologia e Politica (ELSP), a prestigious semi-independent institution of higher learning in the social sciences. Although both institutions were public, the paulista bourgeoisie contributed with funds to their establishment. Thus, it could be said that Brazilian universities, created as part of modernizing projects were born modern. Moreover, from the beginning an emphasis was placed on the teaching of the social sciences. Furthermore, the faculties were partially composed of foreign professors – mostly French at the USP, Americans at the ELSP. Brazilian universities emerged as part of a dense transnational intellectual network. To some extent, they became part of the system of promotion for young French scholars who after spending some time in Brazil would get prestigious positions in France. Such scholars as Fernand Braudel, Claude Lévi-Strauss, Roger Bastide and others spent time in Brazil before becoming famous.

The development of the social sciences was different in Argentina and in Brazil. Although the first chairs in sociology were established in the former country at the end of the nineteenth century, until the end of the Perón regime the teaching of the social sciences in Argentine universities was limited to a number of courses of sociology, psychology and economics taught by part-time faculty who did not do, for the most part, empirical research. The social sciences in Argentina had a strong theoretical orientation and did not offer career possibilities. In general terms, it could be said that the strong politicization of Argentine universities conspired against their modernization. There has been (and still is) a strong relationship between the country's political history and the evolution of the institutions of higher learning. Professors and universities in general have been the first victims of the numerous coups d'etat that devastated the country between 1930 and 1983. However, between the late 1950s and the mid-1960s there was a short period when Argentine universities recovered their dynamism as an engine for the intellectual modernization of the country. It was then when the social science programs (economics, sociology, anthropology and psychology among others) were formally created.

In Brazil, on the contrary, the social sciences were at the center of the university projects aimed at creating a new breed of national and local intellectual and technical elites. This project has been so strong that it survived (unlike in Argentina) the political avatars of the country. Thus "modern" social sciences were institutionalized in Brazilian universities much earlier than in Argentina. In fact, even before the creation of the first universities there was a strong group of social thinkers, socially recognized as such, and they became part of the accepted genealogy constructed by later social scientists. Unlike his Argentine counterparts, Florestan Fernandes, one of the founding fathers of the Brazilian "modern" (i.e., empirical, American-style, institutionalized) social science, had no problems in placing himself at the end of a chain of local "precursors" to the science he contributed to institutionalize. The chain included Euclydes da Cunha, Raymundo Nina Rodrigues, Artur Ramos and, of course, Gilberto Freyre among others.[64] We will see that in Argentina political polarization hindered the construction of a genealogy like this.

When modern social science was introduced in Brazil by foreign scholars in the 1930s psychoanalysis had already been incorporated as a theoretical tool for social analysis in Europe and in the US. Brazil, with its large "primitive population," whose unconscious was supposedly located more on the surface than that of urban "civilized" European people, was seen by foreign scholars as particularly apt for a psychoanalytic approach.[65] Thus, since the early 1930s, the Escola Livre de Sociologia e Política included courses on psychoanalysis taught by Durval Marcondes and Virginia Bicudo (both future founding members of the IPA affiliated Paulista psychoanalytic association). At the USP, on the other hand, formal courses on psychoanalysis would be introduced only in 1954 as part of the training for clinical psychologists. However, since the late 1930s French sociologist Roger Bastide (who would stay in Brazil until the 1950s) introduced psychoanalytic thought in his teaching of sociology. He had developed an interest in themes related to psychoanalysis before going to Brazil. He had written extensively on the mystical life of primitive people, and its manifestation in poetry and the arts. "The primitive" said Bastide in 1931, "is a slave of his emotions." In 1933 he wrote an article titled "Matériaux pour une sociologie du rêve" in which he tried to launch a sociology of dreams.[66] Once in Brazil, Bastide became immediately fascinated by the relationship between different races and cultures, a relationship where he found the key to the originality of Brazilian culture. However, in contrast to Mario de Andrade – with whom Bastide established a fruitful dialogue – according to Bastide, the mixture of races should be understood as a juxtaposition of different cultural elements and not as a "digestive synthesis." Bastide shared with modernists an interest (one could say obsession?) for defining the authenticity of Brazilian cultural production.

In 1941, Bastide wrote his influential article "Psicanálise do cafuné."[67] There, he analyzed from a psychoanalytic-sociological point of view a particular ritual which had been very popular in the patriarchal family during the empire: the caresses administered by female slaves to their mistresses' heads with the apparent purpose of killing lice. After dismissing Freud's *Totem and Taboo* as a fable, Bastide recognized, nonetheless, that sociology and anthropology could benefit from a dialogue with psychoanalysis. Bastide starts his argumentation by discussing the libidinal dimension of the *cafuné*: it generated a pleasure of a sexual kind; in other words, it was a substitute for masturbation. Bastide concludes that *cafuné* was associated with the sexually subordinate place of women in plantation-based patriarchal families, and to the separation between sex and marriage this kind of social formation enabled. Marriages were arranged, but while men had free rein to satisfy their sexual desires with the black female slaves, women were forced to observe a strict sexual morality. Then, *cafuné* was, in Bastide's eyes, a substitute for women's sexual satisfaction, "and thus it had a useful function, because it protected morality."[68] Bastide used cafuné as a case study to conciliate psychoanalysis with social science.

A few years later, Bastide wrote *Sociologie et psychanalyse* based on his lectures at the University of São Paulo.[69] *Sociologie et psychanalyse* was a methodological work in which the author shows that, although there was an incompatibility between classical (e.g., Durkheimian) sociology and traditional psychoanalysis, there was, nonetheless a fertile ground for interdisciplinary work between renovated social sciences and modern versions of psychoanalysis. The last chapter of the volume is a comparative analysis of the relations between blacks and whites in the US and in Brazil. Bastide tries to explain the relations between the two ethnic groups in the US, and particularly certain phenomena such as the existence of a sexual taboo between black men and white woman as the result of a displaced Oedipus complex. In the time of slavery, the white mistress had occupied a maternal position for the blacks. The taboo of the white woman is "identical to the taboo of incest," concludes Bastide. Although slavery had been abolished long ago, the taboo remained.[70] The other side of the Oedipus complex: the hatred of the father, would also be present in the conflicted relations between black and white men. However, and here is where Bastide parts company with "traditional freudians," these sexual elements must be understood in a broader social context and manifest themselves differently according to social class.

In a country like Brazil, on the other hand, where racial lines were not so well defined, things were different. Whereas in the US, blacks (particularly those belonging to the upper social strata) tried to incorporate themselves the best they could into the white society by internalizing white values, in Brazil Bastide found the survival of a large number of African cultural elements, particularly those of religious or folkloric nature. Following the works of Artur Ramos, Bastide wondered whether the survival of myths and rituals of African origins among Brazilian blacks was due to the fact that they continued to fulfill (although in a distorted and modified form) the same kind of unconscious needs that they had fulfilled in Africa, or was it simply the result of the "conservative power of traditions" among religious groups. Unlike the previous generation of doctors and essayists who associated certain African rituals with mental diseases, Bastide maintained that: "The cult of *candomblé*, far from being the root of pathology is, on the contrary, a form of social control of the unconscious. Therefore, it is a factor of psychophysical equilibrium."[71] The trance, therefore should be explained by both psychological and sociological factors. In summary, for Bastide the African rituals that survived in Brazil were a synthesis of collective representations imposed by an ancestral religion and unconscious tendencies which were controlled by tradition. What for earlier Brazilian psychiatrists had been a form of pathology, for Bastide was, in the words of Luis Fernando Dias Duarte, "a balm and compensation for the excesses of civilization."[72]

The institutionalization of social sciences in Argentina took place later than in Brazil. Moreover, whereas in Brazil anthropology and sociology became sciences quickly associated with the construction of a national identity and the understanding of race relations, in Argentine things developed quite differently. There, the institutionalization of sociology was linked to what was seen at the time as "the peronist problem."[73] After the fall of Perón, the intellectuals and professors who had been excluded by the peronist regime became the new leaders of the reformed universities. For them, the modernization of the country had to be associated to its "de-peronization." Thus, they found themselves with the political impossibility to construct the kind of lines of continuity with the past that the Brazilian had done and were doing at the time. The peronist order was conceptualized as a total rupture in the country's history, as a kind of parenthesis that had to be closed forever. While Florestan Fernandes, as it was discussed above, recognized himself as the last link of a genealogical chain, Gino Germani, the Italian emigrée who led the institutionalization of social sciences in Argentina, said that he could not find anything useful in the works of his Argentine predecessors. Moreover anthropology, a science that in Brazil played a central role in the search for a national identity, occupied a marginal place within the Argentine social sciences since it was perceived as a "backward looking" discipline. It was a discipline devoted to the study of those sectors of society that modernity was leaving behind and that, therefore, were bound to disappear.[74]

The construction of an Argentine field of "modern, scientific sociology" as an empirically based science was the work of Germani – who had moved to Argentina escaping from fascism in the late 1930s – and his group of disciples and collaborators. Germani (who would later become one of the most prominent Latin American sociologists) was influenced by American social science, in particular by Talcott Parsons's functionalism. However, Germani's idea of social science was broad and included a combination of certain tendencies within psychoanalytic thought, interpersonal relations psychology and cultural anthropology. Like Bastide Germani also wanted to combine sociology with psychoanalysis; but unlike the French sociologist who cited in his works the latest theoretical findings of Jacques Lacan, Germani's model was precisely what Bastide rejected: American culturalist neo-psychoanalysis promoted by people like Karen Horney, Erich Fromm or Harry Stack Sullivan.[75]

Germani's concerns were also different from those of Brazilian sociologists. While for the latter (including for Bastide) psychoanalysis was a tool to approach the unconscious dimension of Brazilian culture's primitive aspects, Germani's preoccupations were of a more political nature. Germani was worried about the origins and possibilities of totalitarianism. Psychoanalysis (in its "culturalist" version) could provide social sciences with a subjective dimension to analyze the problem of authoritarianism. It is not by chance that Germani translated and wrote the prologue for best-selling *Fear of Freedom* by Erich Fromm, which he considered as an "excellent example of the application of psychoanalysis to historical phenomena," as well as an important contribution to sociological theory.[76] Germani was also the director of two important editorial collections that published works on sociology, social psychology and psychoanalysis. Almost 80 percent of the titles published by one of these collections: "Social Psychology and Sociology" (Editorial Paidós) between 1945 and 1960 were in one way or the other related to psychoanalysis.[77]

In the particular Argentine context the problem of fascism was associated to the experience of Peronism and to the search to alternative paths to political and social modernization, which would constitute one of Germani's main topics of interest. As Alejandro Blanco points out, Germani's use of psychoanalysis was a double innovation. On the one hand, his version of psychoanalysis was antithetical to the orthodox Kleinian version adopted by the Argentine Psychoanalytic Association. On the other hand, his introduction of psychoanalysis to social thought was an innovation in the traditional Argentine sociological establishment and was associated with Germani's interdisciplinary concept of social science. For Germani psychoanalysis had a specific place in social science in moments of crisis produced by fast social change, when explanations of social action required the elucidation of the psychological aspects of human behavior.[78]

Germani was active in the new field of "mental health" as well. After the fall of Perón, the military authorities in government created an "Instituto Nacional de Salud Mental" (National Institute of Mental Health) for the prevention and treatment of mental illness. Unlike the older idea of "Mental Hygiene," "Mental Health" was conceptualized as an interdisciplinary project. The advising committee to the new institution included not only psychiatrists but also social scientists. Gino Germani was among them. However, in spite of Germani's theoretical interest in psychoanalysis and although he wrote several pieces on the intersection between psychoanalysis and the social sciences,[79] it is nonetheless difficult to find much psychoanalysis in Germani's (or his collaborators) empirical work. Nonetheless, as he claimed paraphrasing H.D. Walden's words, psychoanalysis had become a "climate of opinion."

*   *   *   *   *

By the time the "official" psychoanalytic associations were created in the 1940s and 1950s in Buenos Aires, São Paulo and Rio de Janeiro, there was already a public awareness and a social demand for Freud's ideas and techniques. This awareness was the result of a process of reception and circulation of psychoanalysis that was not lineal but multidimensional and that took place at different speeds. This chapter focuses only on some of those dimensions, those more clearly linked to the formulation of national identity.

Studying the reception of systems of ideas raises a number of important theoretical questions. Why certain complex theoretical systems are more successful than others in becoming transnational and at the same time in overflowing their original area of application, turning themselves into a true "systems of beliefs" for broad sectors of very different societies (and the target of the fury of other sectors) is a very hard question to answer which of course fells far beyond the scope of this article. Undoubtely, during the twentieth century, psychoanalysis was one of these successful cases (the same could be said about Marxism and also about Darwinian evolutionism). It is very difficult to understand the past century without taking into consideration the impact that psycho-analysis (and Marxism) had in shaping ideas and visions of the world, both for friends and foes.

What is clear, however, is that the transnational diffusion of a system of thought (psychoanalysis in this case), is made possible and facilitated by certain qualities of that system of thought and of the "recipient" societies. I believe that the process of transnationalization of a doctrine can only be understood by comparing specific patterns of reception. In all cases the process of reception of a certain body of ideas implies an appropriation and simultaneously to a certain extent, a reformulation of the ideas. A system of thought is diffused in a society when it can be fitted into particular and already existent concerns, and when it is compatible with national and social habitus. Here, I compare two cases of early reception (and transnationalization) of psychoanalysis focusing on how psychoanalysis fitted into intellectual programs of construction of national identity and into national habitus. In Argentina, the implantation of psychoanalysis fell victim of political polarization. In Brazil, on the other hand, psychoanalysis was appropriated to provide alternative interpretations to the prevalent views of race relations.

To some extent, the reception of psychoanalysis in both countries was a cross-cultural phenomenon. Prominent Brazilian doctors such as Antonio Austregesilo and Julio Porto Carrero went to Buenos Aires to lecture on psychoanalysis in the 1920s and 1930s. On the other hand, since the 1940s Argentine psychoanalysis had an enormous influence on the development of institutional psychoanalysis in Brazil. A good number of prominent members of the first generation of carioca psychoanalysts were trained in Buenos Aires, and Argentine psychoanalysts traveled regularly to Brazil. A few Argentine analysts moved to Brazil permanently for political or economic reasons during the 1970s. However, as I hope to have shown, the particular path of early reception of psychoanalysis in each country was shaped by cultural and social factors related to broader factors such as the role of intellectuals and their relationship to the state in each country, academic and political traditions and, more generally, national habitus. Psychoanalysis, like many others systems of ideas and beliefs, has a transnational dimension defined by its circulation and diffusion beyond the national frontiers, but also has a national dimension defined by different patterns of reception, appropriation and reinterpretation. Only relatively recently historiography started paying attention to the latter and it is hoped that new comparative studies will contribute to our knowledge of one of the most influential systems of thought of the twentieth century.

Notes

1. In recent years there has been an impressive production of scholarship on the diffusion of psychoanalysis in different countries. What follows is only a small sample of it. For the diffusion of psychoanalysis in France, see Elisabeth Roudinesco, *La bataille de cent ans: L'histoire de la psychanalyse en France*, 2 vols. (Paris: Seuil, 1986) and Sherry Turkle, *Psychoanalytic Politics: Jacques Lacan and Freud's French Revolution* (2nd ed., London: Free Association Books, 1992); for Russia, Martin Miller, *Freud and the Bolsheviks. Psychoanalysis in Imperial Russia and the Soviet Union* (New Haven: Yale University Press, 1998); and Alexander Etkind, *Eros of the Impossible. The History of Psychoanalysis in Russia* (Boulder: Westview, 1997); for the US, see Nathan G. Hale Jr, *The Beginning of Psychoanalysis in the United States, 1876–1917* (New York: Oxford University Press, 1971); and *The Rise and Crisis of Psychoanalysis in the United States, 1917–1985* (New York: Oxford University Press, 1995); for Argentina, Mariano Plotkin, *Freud in the Pampas. The Emergence and Development of a Psychoanalytic Culture in Argentina* (Stanford: Stanford University Press, 2001). For Australia, see Joy Damousi, *Freud in the Antipode. A Cultural History of Psychoanalysis in Australia* (Sydney: UNSW press, 2005).
2. Some authors make the distinction between "freudism" as the loose collection of discourses and practices inspired in Freud's ideas and "psychoanalysis" as a more formalized therapeutic practice. I believe that this distinction in unnecessary. I prefer to use a broad definition of psychoanalysis. See Hugo Vezzetti, *Aventuras de Freud en el país de los Argentinos* (Buenos Aires: Paidós, 1996).
3. See Mariano Plotkin, "Tell Me Your Dreams: Psychoanalysis and Popular Culture in Buenos Aires, 1930–1950," *The Americas*, 55: 4 (1999) 601–629.
4. Elias derives his conception of "habitus" from psychoanalysis: "[Freud] attempted to show the connection between the outcome of the conflict-ridden channelling of drives in a person's development and his or her resulting habitus." Analogous connections could be established for societies. Although I am not sure about the fruitfulness of such analogies, I prefer the (historical) concept of "national habitus" to the (a-historical and essentialist) idea of "national character." See Norbert Elias, *The Germans. Power Struggle and the Development of Habitus in the Nineteenth and Twentieth Centuries* (New York: Columbia U.P., 1996), p. 19 and *passim*.
5. Until the 1920s Argentina was the country that received the largest proportion of European immigrants in terms of its population in the world. Of course, the US received many more immigrants in absolute terms. The census of 1914 revealed that one third of the entire Argentine population was foreign-born.
6. Richard Morse, "The Multiverse of latin American Identity, c. 1920–c.1970," In Bethell, Leslie (ed.), *Ideas and Ideologies in Twentieth Century Latin America* (New York: Cambridge University Press, 1996).
7. For a historical analysis of Argentine elites's dycothomical view of the reality, see Maristella Svampa, *El dilema argentino: civilización o barbarie. De Sarmiento al revisionismo peronista* (Buenos Aires: El cielo por asalto, 1994).
8. On the evolution of ideas about immigration, see Halperin Donghi, Tulio, "¿Para qué la inmigración? Ideología y política inmigratoria en la Argentina

(1810–1914)," Halperin Donghi (ed.), *El espejo de la historia: problemas argentinos y perspectivas hispanoamericanas* (Buenos Aires: Sudamericana, 1987).

9. Lilia Moritz Schwarcz notes that during the Brazilian monarchy a system of titles imported from Europe which included dukes, counts, viscounts, barons, and the like was mixed with indigenous names for those titles. Thus, there was a Viscount of Pirajá; a Viscountess of Tibají; a Baron of Bujurú., etc. See Moritz Schwarcz, Lilia, *As barbas do Imperador. D. Pedro II, um monarca nos trópicos* (São Paulo: Companhia das Letras, 1998).

10. Until the abolition of the slave trade Brazil received more black slaves than any other country in the Americas. On the paradoxical place of liberalism in nineteenth century Brazilian elites' thought, see Roberto Schwarz, "Las ideas fuera de lugar," In Adriana Amante and Florencia Garramuño, (eds.), *Absurdo Brasil. Polémicas en la cultura brasileña* (Buenos Aires: Biblos, 2000).

11. Nineteenth century romantic writers exhalted and idealized the "indian" as a national symbol. Moreover, other writers and critics such as Silvio Romero or Araripe Junior found in "mestizaje" (miscegenation) the origins of Brazilian literature. Later, at the turn of the century, in his enormously popular work *Os Sertões* (1902), Euclydes da Cunha found the bedrock of Brazilian nationality in the mestizo population from the sertão. However, the hegemonic point of view among Brazilian intellectuals held that the whitening of the population was at the same time a precondition and a consequence of modernization. For a discussion on race in literature, see, Roberto Ventura, *Estilo tropical. História cultural e polêmicas literárias no Brasil* (São Paulo: Companhia das letras, 1991).

12. See Thomas Skidmore, *Black into White. Race and Nationality in Brazilian Thought* (2nd ed., Durham: Duke University Press, 1993).

13. See Emilia Viotti da Cosa, "The Myth of Brazilian Racial Democracy," in Viotti da Costa (ed.), *The Brazilian Empire, Myths and History* (Chicago: Chicago University Press, 1985).

14. Many of the most prestigious Brazilian intellectuals active in late nineteenth and early twentieth centuries were either black or mulattoes, including Machado de Assis, the most important Brazilian writer, and founder of the Brazilian Academy of Letters; Raymundo Nina Rodrigues, a prestigious (and racist!) psychiatrist/anthropologist working in Bahia; Juliano Moreira, considered the founding father of Brazilian modern psychiatry, and many others. It is interesting that in most cases their color was forgotten. A Brazilian colleague told me that after working for many years on Moreira's work and reading all his (numerous) biographies she learned that her subject was black, only when she came across a picture of him in an archive. It seems that his prestige had "whitened" him.

15. On the impact of positivism in Latin America, see Hale, Charles, "Political and Social Ideas," In Bethell, Leslie (ed.), *Latin America: Economy and Society, 1870–1930* (New York: Cambridge University Press, 1989). In Brazil, Comtean positivism became, during the early years of the republic, almost a civic religion.

16. Nathan G. Hale, *Freud and the Americans. The Beginnings of Psychoanalysis in the United States, 1876–1917* (New York: Oxford University Press, 1995 [1st ed 1971]), Chapters III and IV.

17. Skidmore, *Black*; Daim Borges, "Puffy, Ugly, Slothful and Inert: Degeneration in Brazilian Social Thought," *Journal of Latin American Studies*, 25: 2 (May 1993) 235–256; See Kristin Ruggiero, *Modernity in the Flesh. Medicine, law and*

Society in Turn-of-the-Century Argentina (Stanford: Stanford University Press, 2004).

18. The image of the (particularly) Italian immigrant as the seed of degeneracy was present in naturalist fiction. See Gabriela Nouzeilles, *Ficciones somáticas. Naturalismo, nacionalismo y políticas médicas del cuerpo (Argentina 1880–1910)* (Rosario: Beatriz Viterbo, 2000).

19. Moreira-Almeida, Alexander, Silva de Almeida, Angélica and Lotufo Neto, Francisco, "History of 'Spiritist Madness' in Brazil," *History of Psychiatry*, 16: 1 (2005) 5–25.

20. This was true even after Freud's *Complete Works (Obras completas)* translated by Spaniard Antonio López Ballesteros became available in Spanish in 1926. It is worth noting that this translation (approved by Freud himself) was the first edition of Freud's complete works in any language. For the evolution of psychoanalysis in France, see Roudinesco, *La bataille.*

21. Germán Greve, "Sobre psicología y psicoterapia de ciertos estados angustiosos," reproduced In Vezzetti, Hugo (ed.), *Freud en Buenos Aires, 1910–1939* (Buenos Aires: Puntosur, 1989), p. 90.

22. See Hugo Vezzetti, *Aventuras de Freud en el país de los argentinos. de José Ingenieros a Enrique Pichon Rivière* (Buenos Aires: Paidós, 1996), pp. 15–18.

23. Raitzin, Alejandro, "La locura y los sueños," *Revista de Criminología, Psiquiatría y Medicina Legal*, 6 (1919), 25–54.

24. Francisco Franco da Rocha, *O pansexualismo na doutrina de Freud* (São Paulo, Typographia Brasil de Rotschild, 1920).

25. See Montechi Valladares de Oliveira, Carmen Lucia., "L'implantation du mouvement psychanalytique à São Paulo" (PhD. Diss, Université Paris 7, 2001), p. 93.

26. Jane Russo, "A Psicanálise no Brasil-Institucionalizacão e Difusão entre o Público Leigo" Paper delivered at the 2006 meeting of the Latin American Studies Association in Puerto Rico (March 15–18, 2006). See also Ana Teresa Venancio and Lázara Carvalhal, "Juliano Moreira: a psiquiatria científica no processo civilizador brasileiro," In, Dias Duarte, Luiz Fernando; Russo, Jane and Venancio, Ana Teresa (eds.), *Psicologizacão no Brasil. Atores e autores* (Rio de Janeiro: Contra Capa Livraria, 2005).

27. See Franco da Rocha, "Do delírio em geral" (opening lecture for the course on Clinical Psychiatry of 1919 at the Medical School of São Paulo), *O Estado de São Paulo* (March 20, 1919), reproduced in *Revista Brasileira de Psicanálise*, 1: 1 (1967), 127–142.

28. However, social thinkers did approach psychoanalysis during the 1960s. See Jorge Balan, *Cuéntame tu vida. Una biografía colectiva del psicoanálisis en la Argentina* (Buenos Aires: Planeta, 1991). For Peronism and culture, see Mariano Plotkin, *Mañana es San Peron. A Cultural History of Peron's Argentina* (Wielmington: Scholarly Resources, 2003). Later, in the 1970s and 1980s the diffusion of the doctrines of Jacques Lacan opened a new space for the interaction between analysits and intellectuals. See also Plotkin, *Freud in the Pampas.*

29. Russo, Jane, "A difusão da psicanálise no Brasil na primeira metade do século XX- Da vanguarda modernista à radio-novela" *Estudos e Pesquisas em Psicologia*, 2: 1 (2002), 53–64.

30. Buarque de Holanda, Sergio, *Raizes do Brasil* (Rio de Janeiro: José Olympo, 1936).

31. Freyre, Gilberto, *Casa grande & senzala. Formação da família brasileira sob o regime de economia patriarcal* (Rio de Janeiro: Maia & Schmidt, 1993).

32. Jane Russo, "A Difusão da Psicanálise no Brasil na Primeira Metade do Século XX-Da Vanguarda Modernista à Radio-Novela," *Estudos e Pesquisas em Psicologia*, 2: 1 (2002), 53–64.

33. Ramos, Artur, *O negro brasileiro* (Recife: Massangana, 1988 [1st ed. 1934]), cit by Alexandre Schreiner, "Uma aventura para amanhã. Artur Ramos e a neuro-higiene infantil na década de 1930," In Dias Duarte, Russo and Venancio (orgs.), *Psicologizacão*, p. 157.

34. Porto Carrero was the author of five volumes on psychoanalysis, translator of Freud's work into Portuguese, director for 5 years of the psychoanalytic clinic of the Brazilian League of Mental Hygiene.

35. Porto-Carrero considered that abortion should be permitted (after a technical commission reviewed each cases) for six reasons: (a) therapeutic; (b) prophilactic; (c) eugenic; (d) moral; (e) aesthetic (in the case of women for whom, like dancers such as Isadora Duncan, the aesthetic of their body constitute the basis of their contribution to society); and (f) professional (women who make a crucial contribution to society and whose pregnancy and motherhood would impair them to fulfill their professional role, such as scientists). He also promoted birth control. See, Porto-Carrero, Júlio Pires, *Psicanalise de uma civilizacão* (Rio de Janeiro: Editora Guanabara, 1935), p. 164 and ff. See also Russo, Jane, "Júlio Porto-Carrero: a psicanálise como instrumento civilizador," In Dias Duarte, Russo and Venancio (orgs.), *Psicologizacão*.

36. Porto-Carrero, Julio, "O carácter do escolar, segundo a psychanalyse" Paper delivered at the First National Conference on Education in Curitiba. Cited in Montechi, Valladares de Oliveira, "Os primeiros tempos."

37. Elisabete Mokrejs, *A Psicanálise no Brasil. As Origens do Pensamento Psicanalítico* (Petrópolis: Vozes, 1993), ch. 4.

38. Porto-Carrero, Júlio P., "Coneito Psychanalytico da pena," In Porto-Carrero (ed.), *Ensaios de Psychanalyse* (Rio de Janeiro: Flores y Mano, 1929), p. 185.

39. Porto-Carrero, Júlio, *Criminologia e psicanálise* (Rio de Janeiro: Flores & Mano, 1932), p. 63, cited in Russo, "Júlio Porto-Carrero," p. 135.

40. For a comparison between Perón's and Vargas's relationship with intellectuals, see Fiorucci, Flavia, "¿Aliados o enemigos? Los intelectuales en los gobiernos de Vargas y Perón" en Rein, Raanan y Rosalie Sitman (comps.), *El primer peronismo. De regreso a los comienzos* (Buenos Aires: Lumiere, 2005).

41. On Brazilian intellectuals, see Miceli, Sergio, *Intelectuais e classe dirigente no Brasil (1920–1945)* (São Paulo: DIFEL, 1979); and Pécaut, Daniel, *Entre le peuple et la nation: Les intellectuels et la politique au Brésil* (Paris: Editions de la Maison des Sciences de l'Homme, 1989).

42. Morse, "The Multiverse," p. 18.

43. However, it should be noted that Modernism was far from a unified movement and developed strong regional and ideological overtones. For an overview, see Cândido, Antonio and Castello, José Aderaldo, *Presenca da literatura brasileira. Modernismo/História e antologia* (Rio de Janeiro: Bertrand Brasil, 2001).

44. In English in the original. De Andrade, Oswald, "Manifesto Antropófago" published in *Revista de Antropofagia*, 1 (May 1926), reproduced in Schwartz, Jorge, *Vanguardas Latino-Americanas. Polêmicas, manifestos e textos críticos* (São Paulo: Edusp, 1995), pp. 142–147.

45. Cândido, Antonio, "A Revolucão de 1930 e a Cultura," *Novos Estudos Cebrap*, 2: 4 (April 1984).

46. Dr. Porto-Carrero, for instance, criticized mondernist precisely for their glorification of what was primitive and infantile. See Porto-Carrero, *Psicanálise*, p. 37.

47. Florencia Garramuño, *Modernidades primitivas. Tango, samba y nación* (Buenos Aires: Fondo de Cultura Económica, 2007).

48. De Andrade, Oswald, "Manifesto Antropófago."

49. Implicit references to Freud's *Totem and Taboo* appear also in fictional works by Modernista authors. See, for instance Mario de Andrade's semi-autobiographical short story of 1942 "O peru de natal" (The Christmas Turkey) in which the ritual eating of a Christmas turkey is openly associated to the conflictive figure of the death father.

50. Cristiana Facchinetti, "Deglutindo Freud: Histórias da digestão do discurso psicanalítico no Brasil" (Ph.D diss. Universidade Federal do Rio de Janeiro, 2001).

51. De Andrade, Mario, "Prefacio Interessantíssimo" to Paulicéia Desvairada (São Paulo, 1922), reproduced in Schwartz, *Vanguardas*, p. 122 and ff.

52. Montechi Valladares de Oliveira, Carmen Lucia, "L'implantation du mouvement psychianalytique à São Paulo," p. 84.

53. See Marialzira Perestrello, "Primeiros econtros com a psicanálise. Os precursores no Brasil (1899–1937)," In Figueira, Sérvulo (ed.), *Efeito Psi* (Rio de Janeiro: Campus, 1988); and Sagawa, Roberto Yutaka, "A psicanálise." Marcondes himself had published literary texts in the modernista magazine *Klaxon*.

54. Boedo and Florida are two streets of Buenos Aires. Florida was one of the most fancy and expensive streets at the time, where the avant-gardists, most of them belonging to prominent families, met. Boedo was considered a middle-class street. The Boedo group was composed mostly by children of immigrants. They were politically leftists, but less daring in aesthetic terms. While the Florida group linked to the journals *Martín Fierro, Proa, Prismas* and others emphasized the autonomy of the arts, the Boedo group wanted to transform literature into a tool for social struggle.

55. For a perceptive analysis of *Martín Fierro*, see Sarlo, Beatriz, *Una modernidad periférica: Buenos Aires 1920 y 1930* (Buenos Aires: Nueva Visión, 1988), ch. 4, and Sarlo, Beatriz and Carlos Altamirano, "Vanguardia y criollismo: la aventura de *Martín Fierro*" in Sarlo and Altamirano, *Ensayos argentinos. De Sarmiento a la vanguardia* (Buenos Aires: 1997), pp. 211–260.

56. See "Manifiesto de *Martín Fierro*. Periódico quincenal de arte y crítica libre" in *Revista Martín Fierro 1924–1927. Edición Facsimilar* (Buenos Aires: Fondo Nacional de las Artes, 1995), p. XVI.

57. See Jorge Schwatz, *Vanguarda e cosmopolitismo na década de 20. Oliverio Girondo e Oswald de Andrade* (São Paulo: Editora Perspectiva, 1983), pp. 132–133, 143. Girondo was a conspicuous member of the *Martín Fierro* group: he wrote the

Manifesto. His books *Veinte poemas para ser leídos en el tranvía* (1922), and *Calcomanías* (1925) include sexually charged avant-garde poems.

58. Ibid., p. 192.

59. The problem of genealogy appears conspicuously in Borges's fiction. See for instance his (probably best) short story "El Sur' which starts with the dual genealogy (immigrant and local) of Juan Dahlmann, the main character. References to recent immigrants and the use of language can be found in other stories like "El Aleph" and the saga of Honorio Bustos Domecq (collective pseudonim of J. L. Borges and Adolfo Bioy Casares).

60. Text by Roberto Mariani included in Vignale, Pedro Juan and Tiempo, César, *Exposición de la actual poesía argentina* (Buenos Aires: Minerva, 1927), cit. by Prieto, Adolfo, "Roberto Arlt: Los siete locos. Los lanzallamas," In Arlt, Roberto, *Los siete locos. Los lanzallamas* (Caracas: Biblioteca Ayacucho, 1978), p. XVI.

61. See, Beatriz Sarlo, *La imaginación técnica. Sueños modernos de la cultura argentina* (Buenos Aires: Nueva Visión, 1992).

62. Plotkin, "Tell Me your Dreams."

63. *Los siete locos* and *Los lanzallamas* tell the story of a group of bizarre people who try (and fail) to organize a badly defined revolutionary secret society financed by a chain of brothels. In order to carry out their plan, the characters plan to utilize modern weaponry based on scientific principles. The books include mathematical formulas and diagrams and plans of the gas producing plant, etc.

64. Florestan Fernandes, "Desenvolvimento histórico-social da sociologia no Brasil" in Fernandes, *A sociologia no Brasil. Contribução para o estudo de sua formação e desenvolvimento* (Petrópolis: Voces, 1977), p. 35 and ff.

65. Similarly, in 1929 Hungarian anthropologist-psychoanalyst Geza Roheim took an extensive trip to Australia to carry out a psychoanalytic-anthropological study of native groups. See Damousi, *Freud in the Antipode*, pp. 93–96. See also Damousi's contribution to this volume.

66. Roger Bastide, "Matériaux pour une sociologie du rêve" in *Revue Internationale de Sociologie* 41: 11–12 (1933). An excellent analysis of Bastide's intellectual trajectory can be found in Peixoto, Fernanda Arêas, *Diálogos Brasileiros. Uma análise da obra de Roger Bastide* (São Paulo: Edusp, 2000).

67. Originally published in 1941, in a book that bears the same title, it was reproduced in Bastide, *Sociologia do folclore brasileiro* (São Paulo: Anhambí, 1959).

68. Ibid., p. 320.

69. Roger Bastide, *Sociologie et psychanalyse* (Paris: PUF, 1950). Bastide mentions in the prologue that the book originated in the interest that his Brazilian students had shown for psychoanalysis back in the 1930s.

70. Ibid., p. 242.

71. Ibid., p. 251.

72. Dias Duarte, "Em busca do castelo interior. Roger Bastide e a psicologização do Brasil," In Dias Duarte, Russo and Venancio (eds.), *Psicologização* (Rio de Janeiro: Contra Capa Livraria, 2005), pp. 167–182.

73. See Neiburg, Federico, *Los intelectuales y la invención del peronismo. Estudios de antropología social y cultural* (Buenos Aires: Alianza, 1998).

74. The fact that among the "founding fathers" of Argentine modern anthropology were a few Italians and an Austrian formerly associated to Fascism and Nazism did not contribute to promote the image of anthropology as a "forward-lookingscience." Austrian Dr. Oswald Menghin, one of the leaders of Argentine anthropology had been President of the University of Vienna during the Nazi regime and served for a short period of time as Minister of Education during the government of Arthur Seyss-Inquart.

75. A similar approach to psychoanalysis would be made by Brazilian sociologist Florestan Fernandes, disciple and successor of Bastide in the chair of sociology. Fernandes is usually considered the father of "scientific sociology in Brazil" and shared with Germani a devotion to American functionalist sociology (combined in his case with some Marxism). On Fernandes's culturalist approach to psychoanalysis, see, Fernandes, "Pcianálise e sociologia" Paper delivered on May 6, 1956 in homage to Freud in his centenary organized by the Rio de Janeiro Psychoanalytic Society at the Ministry of Education.

76. Gino Germani, "Prefacio a la edición castellana," Erich Fromm, *El miedo a la libertad* (Buenos Aires: Paidós, 1947), pp. 9–11, cit by Alejandro Blanco, *Razón y modernidad. Gino Germani y la sociología Argentina* (Buenos Aires: Siglo XXI, 2006), p. 127. My discussion on Germani is mostly based on this work and on other papers by Blanco: Alejandro Blanco, "Gino Germani: las ciencias del hombre y el proyecto de una voluntad política ilustrada," *Punto de Vista 62* (Dic. 1998) 25–38; Hugo Vezzetti, "Las ciencias sociales y el campo de la salud mental en la década del sesenta," *Punto de Vista*, 54 (April 1995) 29–33. On Germani, see Alejandro Blanco, "La sociología: Una profesión en disputa," In Federico Neiburg and Mariano Plotkin (eds), *Intelectuales y expertos. La constitución del conocimiento social en la Argentina* (Buenos Aires: Paidós, 2004); Ana Alejandra Germani, *Gino Germani, del antifascismo a la sociología* (Buenos Aires: Taurus, 2004).

77. Blanco, *Razón*, p. 126.

78. See Gino Germani, "Psicoanálisis y sociología: un problema de método" originally published as an introduction to the Spanish translation of *Psicología y sociología* by Walter Hollitscher (Buenos Aires: Paidós, 1951), reproduced in Alejandro Blanco (ed.), *Gino Germani: La renovación intelectual de la sociología* (Bernal: Universidad Nacional de Quilmes, 2007), p. 124.

79. See, for instance, Gino Germani, "El psicoanálisis y las ciencias del hombre" in *Revista de la Universidad*, La Plata, 3 (1956); and Germani, "Sociología, relaciones humanas y psiquiatría," *Revista de la Universidad de Buenos Aires* 1: 1 (1956) 139–144; Germani, "Psicoanálisis."

Section 4
Challenging Centre and Periphery

7
Paris–London–Buenos Aires: The Adventures of Kleinian Psychoanalysis between Europe and South America

Alejandro Dagfal

Melanie Klein (1882–1960) was the first analyst who managed to construct an original system of thought, contesting many Freudian principles, without being forced to leave the psychoanalytic movement. This chapter deals with the strange ways in which her theories were easily transmitted from London to Buenos Aires, in the late 1940s and the early 1950s, whereas she had to wait until 1959 for her first book to be translated into French. Nevertheless, before this date, a curious trilateral circulation of Kleinian ideas – from London to Paris via Buenos Aires – had already been possible, thanks to the action of certain analysts, like Enrique Pichon Rivière, Ángel Garma and Willy Baranger. They were European immigrants installed in South America, who succeeded to build unexpected bridges between the Old and the New Worlds. In this process, as we shall see, the role played by their wives – who became analysts as well – was also very significant.

It would be quite difficult to analyse the construction of psychoanalytic discourse in Argentina, during the second half of the twentieth century, without studying the local reception of Kleinian ideas. 'Kleinism' was one of the major theoretical ingredients – if not the most important – in all psychological discourses produced in that country during the 1950s and 1960s. Nevertheless, in this chapter, we will not limit the scope of our study to the constitution of the Argentine context. We shall focus instead on the migrations of people and the circulation of ideas that permitted the implantation of Kleinism, not only in Argentina, but also, from Argentina, in other countries, such as France.

The history of Melanie Klein herself is a very 'transnational' one. The itinerary of the first psychoanalyst who openly disagreed with some of

Freud's ideas, forging her own theories without leading neither a schism nor a dissident movement, involved several migrations. Born in Vienna, in 1882, she settled in Budapest, in 1910. There she became interested in psychoanalysis, and started a 'talking cure' with Sandor Ferenczi, in 1914. Motivated by the latter, without studying medicine, she consecrated to child analysis, beginning with her own son, Erich. In 1921, fleeing from the anti-Semitism that was ravaging Hungary, the young Melanie moved to Berlin, where she joined the *Deutsche Psychoanalytische Gesellschaft* and began a second analysis with Karl Abraham, in 1924.

Supported by Abraham (from whom she borrowed some important ideas), that very year, she placed herself in the middle of the first debates on the theory and the technique of child analysis (as well as on psychoanalysis in general). These controversies were going to oppose her to Anna Freud, the other renowned child analyst of the time, for the rest of her life. While Freud's daughter proposed a type of treatment that was almost 'educational', centred in the Ego and the defence mechanisms, her opponent analysed children's play and interpreted transference in the light of infantile unconscious fantasies. If the former considered child treatment as radically different from adult treatment (because of the unfinished character of the Oedipal complex), the latter thought it fit to analyse even the aggressive tendencies dominating the baby from the very beginning of life (all the more so since she considered that the existence of the Superego was previous to the Oedipal complex).

In 1925, after Abraham's death, Ernest Jones, the founder of the British Psychoanalytic Society (BPS), invited Klein to move to London, which she ended up accepting the following year. Jones was attracted to Klein because of several theoretical subjects and common interests. Like his guest, he believed in the existence of a specifically feminine sexuality and in the existence of an early Superego. Moreover, he had great curiosity for child analysis. But his interest was not only theoretical, considering that, as soon as Klein arrived in London, Jones gave her the difficult task to analyse his two sons and his own wife. In 1927 – one year after the arrival of the Viennese lady – ,Sandor Ferenczi (her former analyst in Budapest) showed his concern about the Kleinian influence on the English group as a whole. That is why he wrote to Freud: "Jones is not only adopting Frau Klein's method but also her personal relationships with the Berlin group. Apart from the scientific value of her work, I find it an influence directed at Vienna."[1] From that moment on, the dispute between Anna Freud and Melanie Klein would become a conflict between 'the Freuds' and the 'English school', that is to say, between Vienna and London.

Regarding the international context, it seems rather evident that, in the 1920s, after the fall of the Austrian-Hungarian Empire, the capital of the psychoanalytic kingdom had moved from Vienna to Berlin, and was even moving from Berlin to London. This Anglo-Saxon supremacy was yet to be strengthened in the 1930s, due to the rise of Nazism. And this would be confirmed not only by the fact that the Freuds ended up moving to England, but also by the successive re-elections of Ernest Jones as president of the International Psychoanalytic Association (IPA), between 1934 and 1949 (after a first term going from 1920 to 1924). In 1935, before arriving in Berlin, with the intention of implementing his policy of 'rescuing psychoanalysis' in Germany, Jones stopped by Vienna. There, he clearly laid out some of the theoretical differences that were going to divide the psychoanalytic movement during the decade to come: 'For some years now it has been apparent that many analysts in London do not see eye to eye with their colleagues in Vienna on a number of important topics'.[2]

He referred in particular to the early development of sexuality ("especially in the female"), the genesis of the Superego and its relation to the Oedipus complex, the technique of child analysis and the conception of the death drive. It would be difficult to deny that, from 1932 on, the publication of Melanie Klein's first book had contributed to further separate the ideas of these two groups. Indeed, *The psycho-analysis of children*, published almost simultaneously in English and German, posed original theoretical questions, concerning a realm that was almost unexplored until then.[3] Its author, relentlessly backed up by Jones, had already obtained a privileged position in England that would go unchallenged until the arrival of the Freuds, in 1938.

A few years later, in 1943, right after the foundation of the Argentine Psychoanalytic Association (APA), Jones would make the following comments, in a friendly message addressed to the new institution:

The knowledge of German, still desirable, was once an essential tool for the purpose of international links related to our work, but today it is yielding its place to English. We may expect that the growing political collaboration between the Spanish speaking countries and the English speaking ones will also be followed by a close partnership in our scientific work.[4]

Apparently, the complete failure of the initiative undertaken by Jones, in 1935, to 'save' psychoanalysis in Nazi Germany, was followed during the war by a political programme of 'degermanisation' and 'Angliciza-tion'. In 1929, German was the mother tongue of 90 per cent of IPA members. Twenty years later, the proportions would be quite different. In 1949, out of the 800 members of the IPA, 450 were from the USA, and 122 came from the United Kingdom.[5] This means that more than two thirds lived in Anglophone countries. At this point, where the hege-mony of the English language had already been established, we should also state that, by the 1940s, the American Psychoanalytic Association (APsaA) – founded in 1911 by Ernest Jones himself – had become even more important than her British homologue. In fact, the 'anglophoni-sation process' of Freudism had been accentuated by the rise of the USA as a world power, which favoured the choice of many European analysts trying to emigrate. In this general context, we may ask ourselves why the first psychoanalytic institution created in Argentina, in December 1942, ended up having a British orientation and not a North American one.

The founders of the Argentine Psychoanalytic Association

We are not going to tell the story of the APA, since it has already been studied by several authors.[6] Nevertheless, we are going to present very rapidly the itineraries of some of its founding members, to better understand the complexity of their theoretical backgrounds.

Enrique Pichon Rivière (1907–1977) is, without any doubt, one of the most fascinating characters in Rio de la Plata's 'psy history'. An unortho-dox psychiatrist, he would become a psychoanalyst, before turning into a social psychologist. In his restless life, he was also sportsman, journalist and art critic, without forgetting that he was a *bon vivant*. Born in Geneva, his French parents emigrated to Argentina in 1910, settling in the Northeast region, where 'the little Frenchman', as they called him, grew up in contact with the aborigine culture. After studying medicine in Rosario, he specialized in psychiatry, without abandoning his avant-garde interests. Near the end of the 1930s, as he worked in an asylum, his innovative spirit made him take an interest in psycho-analysis. By this time, he began to see Arnaldo Rascovsky (1907–1995), a paediatrician who was a friend of his brother-in-law. Rascovsky, the son of Russian Jewish immigrants, had a particular curiosity for psychoso-matics. He was very interested in psychoanalysis as well. He held a good position in the Buenos Aires Children's Hospital and had a large private clientele. Together, Pichon Rivière and Rascovsky were going to function as the 'local core' that made the foundation of a national movement possible.

To achieve this goal, they had yet to wait for the arrival of Ángel Garma (1904–1993), a Spanish psychiatrist fleeing the civil war, Celes Cárcamo (1903–1990), an Argentine physician who had specialized in neuropsychiatry in France, and Marie Langer (1910–1987), an Austrian doctor who had first fled from Nazism to later escape from Francisco Franco's dictatorship. They had all undergone analytic treatment in Europe, including a training psychoanalysis. Garma had been trained in Berlin, were he had been analysed by Theodor Reik. Cárcamo became member of the *Société Psychanalytique de Paris*, after passing by Paul Schiff's couch. Finally, Marie Langer obtained her training in Vienna, where she was analysed by Richard Sterba. This heterogeneous medical group, with the help of some less conspicuous members, founded the psychoanalytic association that was destined to be the most important in Latin America, in December 1942.

In the foundation of this association, the role played by immigration was not a minor one. Most of the 'authorized' analysts came from Europe, and they would have never arrived in the Pampas had it not been for the spread of Nazism and Fascism. Otherwise, even Cárcamo, the only Argentine analyst trained abroad, would have probably stayed in Paris, where he had already married a French woman. As for the 'local' members, as we have seen, they were in fact the offspring of immigrants, and their adoption of a national identity was therefore rather recent. In a sense, considering the ways by which 'official' psychoanalysis arrived, Argentina was very similar to the US. In the beginning, both countries largely benefited from the analytic Diaspora caused by the European situation, and this facilitated the organization of institutional movements.

Nevertheless, as Plotkin has pointed out, the Argentine case was very original, if we take into account the fact that, very rapidly, psychoanalysis became a 'native discipline', integrating itself into the national cultural landscape.[7] As opposed to the US, those who practised psychoanalysis in Buenos Aires were not identified as foreigners and Jews, basically, because after a few years, they were neither one nor the other. As a matter of fact, only Rascovsky and Langer where Jews, and then, only Langer fit the stereotype of the analyst forged in the US, speaking with a strong German accent.[8] If there were a few prominent foreigners within the association, the same was true for almost any given institution at that time. In any case, the APA was a good example of what happened in the society as a whole.

Moreover, in the absence of an 'absolute leader', the institution functioned in a rather horizontal manner. Although Garma was the eldest

analyst, Cárcamo had similar qualifications. In the 1940s, they alter-
nated as presidents of the association. In the 1950s, when Cárcamo
moved away from institutional life, it was Pichon Rivière's turn to
counter-balance the power of Garma. Analysed by the latter, Pichon
Rivière rapidly conquered an independent position, backed up by his
contacts in the psychiatric field and by his success as a training analyst with the candidates of the second generation. Besides, as Balán has
shown, at the beginning, Garma had needed Pichon Rivière and Rascovsky to become a training analyst himself, for they were his first patients
in analytic training. At the same time, as the leaders of the 'local core',
they sent quite a few clients to the Spaniard, and obtained the necessary funding for the development of the association. Finally, when
Garma needed a second analysis after breaking up with his first wife,
he chose Marie Langer, the youngest member of the founding group.
Thus, even if the APA was by no means a democracy, thanks to these
crossed-dependencies, power was fairly well distributed. It is our hypothesis that this lack of strong personal leadership was soon compensated
by the presence of a strong allegiance to a single theory: the Kleinian
one. But the question pertaining to the causes of this particular choice
still remains unanswered.

The Kleinian choice

In order to study the early reception of Kleinism in the Argentine movement, it is very useful to take a quick look at the first numbers of the
association's organ, the *Revista de Psicoanálisis*, created in 1943. Based
on the analysis of the subjects of study and considering the authors that
were translated in the first number of that journal, it would be hard to
affirm that Klein's ideas were prevalent at the very beginning. On the
contrary, what appears in the first place is the multiplicity of theoretical
references, going from Karl Menninger to Franz Alexander, passing by
Anna Freud and Melanie Klein. Aside from Sigmund Freud, no author
had his place guaranteed, in a context where the main interest seemed
to be psychosomatics. In the following two or three years, however,
Garma, Pichon Rivière, Langer and, to a lesser extent, Rascovsky, would
all adopt some basic Kleinian principles, and show it in their writings
published in the *Revista*.[9] With the exception of Cárcamo, by 1950, the
most important founding members understood psychoanalysis in the
light of a Kleinian matrix underlining, among other things, the importance of the initial relationship with the mother, the early constitution
of the Superego and the significance of fantasies and internal objects.[10]

There is no easy explanation for this 'collective conversion'. Everything shows that Garma and Pichon Rivière were the introducers of Melanie Klein in a *theoretical* level, whereas their wives were the first to use her ideas in a more *practical* level. They were the first members of the APA to devote themselves to child analysis. Marie Langer fell somewhere in between these two groups. She was as interested in theoretical matters (using the works of Klein even to interpret social phenomena) as in child analysis and the problems of motherhood.[11] In 1942, she read for the first time *Die Psychoanalyse des Kindes*, which she borrowed from Bela Székely, another European émigré who had been close to the founding group.[12] That allowed her to help Arminda Aberastury, who was at the time translating into Spanish chapter eight of the English version (which would be included in the first issue of the *Revista de Psicoanálisis*).[13] She was also considering translating the whole book, which would not be published until 1948, by El Ateneo.

Aberastury (1910–1972), the first wife of Pichon Rivière, was a pedagogy teacher. Being a member of a well off family of Catalan origins, she shared her husband's passion for philosophy and the arts. In 1937, she took an interest in child analysis, supposedly stemming from the contact she had in her husbands' waiting room with the patients' children.[14] She then began to treat children in the asylum Las Mercedes, expanding her knowledge on the subject by reading Anna Freud and Sophie Morgenstern.[15] By 1942, Aberastury's analyst, Ángel Garma, seeing her interest in child analysis, finally oriented her towards the Kleinian works. In 1945, she started a personal correspondence directly with Melanie Klein and a 'postal supervision' that would last until 1957.

> My work was definitely influenced by the reading of *The psychoanalysis of children*. I was discouraged by the obscurity of many paragraphs and by the richness that I imagined behind this obscurity. When I wrote her a letter [to Melanie Klein] pointing out some of my difficulties and announcing that I had undertaken the translation of her book, I received this response, dated on April 27, 1945, that answered three questions on the relationship with the parents, the handling of play material and the payment of honoraries.[16]

It is clear that the first exchanges between these two women took place at a rather technical level, and were motivated by practical concerns on the ways to proceed in child analysis. That is why Aberastury could then combine Klein's techniques with those proposed by Anna Freud, as she pleased. Still in 1945, as her translation of the book was progressing, she realized that she needed an English speaker to correct the text. Thus, Ángel Garma put her in contact with Elizabeth Goode (1918–2003), an Argentine singer from a British family, who also gave private English lessons for a living (she had several students in the Rascovsky family). Moreover, she had helped Garma with the translation of some papers that he presented in New York in 1943.[17] She had already been in analysis with Marie Langer for a year. Goode would very soon become Argentina's second child analyst, as well as Ángel Garma's second wife.

The particular meanings of the Kleinian choice

As we have seen, the hegemony of Kleinism in the Argentine asso-ciation, more than an initial datum, was the result of a progressive historical construction. At first glance, everything shows that the the-oretical positions of the male members were very eclectic, with different nuances. The women, for their part, even if they were well informed about the most up-to-date debates, were focused on more precise inter-ests, like childhood and motherhood. In this respect, the matter of gender is not a minor one. In the first place, if child analysis was consid-ered a 'women's occupation', in the 1940s, the BPS had almost become a 'women's institution'. After Freud's death in London, in 1939, his legacy was mainly being disputed by two women: his own daughter and Melanie Klein.

In this succession quarrel, Anna Freud evoked the authority of the father. But she also embodied the continuity of the *status quo*, particu-larly regarding the phallocentric conception of woman. Furthermore, in a very schematic way, we could say that she represented the values of dis-cipline and education, aligning herself more with repression rather than the repressed. She seemed to be closer to the normative world (generally associated with masculinity) than to the expressive realm (usually linked with femininity). Melanie Klein, on the contrary, presented herself as the innovative and audacious woman who had to keep psychoanalysis alive. Declaring to be loyal to 'the best Freud' (the one of the 1920s), sur-rounded by a women guard (Joan Rivière, Paula Heimann, Hanna Segal, etc.), she would risk going farther into the heart of the unconscious, claiming to have arrived where the Viennese master had not dared. As a woman, she could better understand the marvels and the horrors of the archaic relationship between mother and child. She was in the posi-tion to grasp the essence of femininity beyond the phallus, and even to venture herself into the bowels of psychosis.

Thus, it is easy to understand how Marie Langer – a former socialist activist, who had participated as a volunteer in the Spanish Civil War and would later become a feminist – could come to prefer Melanie Klein to Anna Freud. Klein not only seemed more attractive as a theoretical source, but for the women of the APA she would also serve as a professional identification model, not exempt of political connotations. She was likely to be seen as a self fulfilled woman who had had the nerve to confront the psychoanalytic establishment. If the wives of Pichon Rivière and Garma (Arminda Aberastury and Elizabeth Goode, respectively), ended up practicing child analysis from a perspective that was more or less Kleinian, it was partly because of the reasons we have just expounded. The choice made by their husbands, even though it was made earlier, seems more difficult to explain.

The case of Garma is suitable for different hypothesis that are not mutually exclusive. In any case, his stay in Berlin, when Abraham had just died and the footprints of Melanie Klein were still fresh, seems to have been crucial. Like Paula Heimann, he was analysed by Theodor Reik, was supervised by Otto Fenichel and followed Franz Alexander's courses, among others. He already showed a considerable dose of self-determination. In 1931, for example, in a paper he presented for acceptance at the *Berliner Psychoanalytisches Institut* (BPI), he dared challenge some of Freud's theories on psychosis, especially those regarding the abandon of reality and repression.[18] Later on, while he was still in France, Garma would state that psychoanalysis was divided in three theoretical trends: the Reichian one, with a social vocation, the Kleinian one, focused on the unconscious fantasies, and the Annafreudian one, halfway between the other two. At the time, he seemed more attracted by this last trend, although this would shortly change. In 1945, as we have mentioned, he would review an article written by his friend and former mate at the BPS, Paula Heimman, who was already Melanie Klein's right hand.

> The article is very worthy in order to evaluate the therapeutic process in psychoanalysis. For a better understanding of the psychological mechanisms, it may have been convenient that the author had gone further into the genesis of the patient's aggressiveness that was directed towards her internal objects.[19]

Thus, Garma, using expressions such as 'bad object' and 'introjected parents', was already expressing himself in terms that were perfectly Kleinian, which would only be more evident in the future. If Paula

Heimann, who had received the same analytical training in Berlin, and had been analysed by the same person, could become fiercely Kleinian, did it not make sense that Garma could follow a similar path? In their common experience in Berlin and in their analysis with Reik – who had been analysed at his turn by Abraham – was there not anything that related them to Melanie Klein? Had not Abraham himself enunciated the theoretical principles that were going to orient the Berlin School? Had he not underlined the pre-genital stages in the development of the libido, pointing out the importance of oral and anal sadism? Had he not been the first to dare argue with Freud about non phallic sexuality in women? Had he not shown an interest – as Garma would do later – in the relation between the dreams and the myths, as well as in the maniac-depressive psychosis? If Klein, who had been analysed by Abraham, was so grateful to him that she dedicated her capital book to him, is it not understandable that Garma, trained in the school founded by Abraham (whom he admired) could have several coincidences with her points of view?[20] Everything shows that they shared some of the Abrahamian mandates, which could be synthesized in a couple of sentences: 'always go to the earliest stages of life and to the deepest layers of the unconscious'; 'with Freud, but beyond Freud'.[21]

As for Pichon Rivière, the influence exerted by Garma cannot be more obvious, since he was his first analyst and trainer. Thus, he already had quite a few reasons to feel close to Kleinian ideas. But he had some others of his own. Well before his 'encounter' with Kleinism, Pichon Rivière had shown an interest for madness, becoming a psychiatrist. This professional career, however, did not prevent him from being seduced by a certain marginality, related to tango and to the cabarets that proliferated in Buenos Aires in the 1930s. In this particular context, he was also captivated by the poetry of Rimbaud and the count of Lautréamont. The latter, especially, became an identification model for him, and almost an obsession. Born in Uruguay in 1846, during the siege of Montevideo, Isidore Ducasse (his real name) was an inspirer of the surrealists. He succeeded, better than anyone previously, in forging a 'modern' poetic approach to the sinister and tragic. His *Chants de Maldoror*, published in 1869, were a sarcastic celebration of evil and death, which was later recovered by André Breton and his group. According to the testimony of Pichon Rivière:

My family, as Lautréamont's, was French; both of us lived in an unknown world. More precisely, my childhood, as his, was a great odyssey...In addition, hadn't I been marked, like Lautréamont, by the 'phantoms' of mystery and sadness? [22]

In fact, considering Pichon Rivière's theoretic production, fascination for mystery and sadness were probably the dominating feelings throughout his life. He attributed this to his early exposure to the mythical world and to the magic thought of the Guarani Indians, whose culture placed death as an organizing value. This belief had some empirical foundation, since he had been raised in the provinces of Chaco and Corrientes, a region of tropical forests, full of wild animals and peopled by some aborigines. In this setting, he learned first to speak French, then Guarani, and finally Spanish, in a rural school. According to all of his later testimonies, he was profoundly marked by this childhood as a little immigrant immersed in very strange culture. In his adolescence, he wrote poems dedicated to death, like 'Rencontre avec la femme noire' [Encounter with the black lady].[23] Thus, his interest for Lautréamont, which would never abandon him, seemed to be an expression of his own existential search, of his long fight with the phantoms that haunted a curious little child who had grown up in the crossroads of two opposite worlds.

In any case, it is likely that Pichon Rivière could find in Kleinian theory (especially in her conception of fantasy and sublimation) a tool that was particularly appropriate to integrate his literary and aesthetic interests with his work as a psychiatrist. As opposed to the surrealists, who made use of poetry and psychoanalysis to question the normative ideals of medicine, Pichon Rivière only used them – at least at the beginning – to enrich his medical research. For example, in the early 1940s, he placed epilepsy as the paradigm for psychosis and for mental illness in general.[24] And he did so with the support of several literary references. For Freud, Dostoievsky had been the *princeps* case of epilepsy; for Pichon Rivière, it would be Gustave Flaubert, whom he quoted directly in French. This choice of epilepsy implied several consequences, if we compare it to hysteria, the cornerstone of psychoanalysis, according to Freud and to the surrealists as well. Hysteria underlined the problem of femininity and of a sexual body. Pichonian epilepsy, however, raised the question, rather masculine, of a body tortured by aggressiveness, guilt and the death drive.[25]

The French psychiatrist interpreted the epileptic crisis as a symptomatic formation or, more precisely, as the result of a very primary conflict between the deadly hatred of a sadist Superego and the weak defences of a masochist Ego. Thus, the convulsions represented situations close to death, serving as an outlet for the violence of Superego's energy. This change of perspective from *Eros* to *Thanatos*, this sort of 'desexualisation' of theory did not limit itself to the scope of psychopathology. In the realm of art, Pichon Rivière considered creation as an escape to the ominous, as a sublimation of the death drive (already understood as a destructive force, from a strictly Kleinian position). For him, death was even more impossible to put up with than sex. In this sense, underlying every piece of art there was a fundamental angst, an empty space, without representations, that the artistic creation tried to fill. That was completely compatible with his fascination for the works of Lautréamont and other 'damned poets'.

In short, for several reasons, the Pichonian conception of subjectivity, as much as Garma's, involved an ego that was subjugated, at the mercy of the unconscious and the Superego. It had to do with a traumatized and guilty subject, much closer to the ideas of Klein and Abraham than to the autonomist postulates of Anna Freud and the American Freudians.

Paris–Buenos Aires, Buenos Aires–London, Buenos Aires–Paris, London–Paris

So far, in the triangle Paris–London–Buenos Aires, we have basically examined Kleinism in the vector going from London to Buenos Aires. We shall now take other segments into account, albeit briefly. To consider the vector Paris–Buenos Aires, we cannot underestimate the importance of the French influence on Argentine culture. In psychiatry and psychology, in particular, this influence was even more patent.[26] Hence, we must remember that, in the 1930s, Garma and Cárcamo stayed in Paris for a few years, before joining the other founding members of the APA in Buenos Aires. When they finally arrived, both already married to French women, they had quite a few acquaintances in the Parisian analytic milieu, like Daniel Lagache and René Laforgue. As for Pichon Rivière, he had no need to go to Paris to be initiated in the French thought.

In 1946, a young French philosopher attended Pichon's conferences about the Count of Lautréamont, which took place in Buenos Aires at the *Institut français d'études supérieures*. His name was Willy Baranger, and he did not know that this encounter was about to change the course of his life.[27] Born in Algeria, in 1922, he grew up in Paris. In 1939, because of the war, he moved to the South. He studied philosophy in Toulouse, where he obtained the 'agrégation', in 1944. In 1946, at the age of 24, he was appointed professor of philosophy at the French Institute of Buenos

Aires, where he succeeded Roger Caillois. At the time, this institute was very renowned, and there Baranger hoped to share his knowledge about Jean-Paul Sartre and existentialism. A couple of months after his arrival, the encounter with Pichon Rivière produced a slight change of plans. A year later, Baranger told the story in a letter he addressed to Daniel Lagache, who was one of the more promising figures of the second generation of French analysts:

> Finding myself in Buenos Aires, I have taken advantage of the presence in this city of an active and serious psychoanalytic movement, to start a training analysis with Doctor Pichon-Rivière. This possibility, as you know, did not exist in the provincial towns where I used to teach in France. Now I follow the training courses of the Institute of Psychoanalysis – particularly those of Dr Garma – and work regularly at the Hospicio de las Mercedes [...].

> Maybe you will accept to help me in a completely different level: I have recently become a member of the editorial board of the psychoanalytic journal of Buenos Aires. Dr Arnaldo Rascovsky, the director of the journal, has charged me to summarize all the psychoanalytic and psychological documentation in the French language (books and journals). All the abstracts that I write, or that are sent to me, shall appear in the Argentine journal. On the other hand, the journal shall publish the record of the activities of the *Société française de Psychanalyse* [...].

> If this interests you, I will send you some numbers of the Argentine journal, so that you can see that it has nothing to envy to the best ones in the world. Likewise, it would be possible to publish French authors' articles in the journal (it currently publishes numerous articles of foreign authors, mainly North Americans – and a few articles excerpted from the *Revue française*).[28]

Very quickly, in spite of his youth, Baranger not only obtained an interesting position in the APA, but, being a polyglot, he also found himself in the centre of an ambitious editorial project, as an international ambassador with multiple missions. His interlocutor, Daniel Lagache, had just been appointed professor of psychology at the Sorbonne. Along with Jacques Lacan, he was Melanie Klein's closest ally in the *Société Psychanalytique de Paris*. In the 1950s, after a series of misunderstandings between Lacan and Klein, Lagache would stay as her only reliable contact in France, even if in a theoretical level his ideas were much more compatible with Anna Freud's.[29] As we will see, Lagache and Baranger would be two important characters in the adventures of Kleinism in the years to come, particularly concerning the vectors going to Paris from London and Buenos Aires.

In 1949, the IPA organized its first congress after the war in Zurich. In that congress, Ángel Garma and his wife, Elizabeth Goode, approached Melanie Klein for the first time. Thanks to the mediation of Paula Heimann, a common friend, and the fact that Goode was fully bilingual, the Garmas were able to establish a fluid relationship with the famous analyst.[30] Hence, in January 1952, during the Argentine summer, they would spend a few weeks in London, being supervised by her on a daily basis. According to Betty Goode, the relationship between Klein and her husband was not always very smooth. On one occasion, when Klein did not agree with Garma's point of view, she is supposed to have said: 'The problem is that you are not Kleinian enough'. 'It is true – would answer the Spaniard –, because I am Garmian'.[31] Beyond the anecdote, it is important to note that Garma, as all the APA's first generation analysts who became Kleinian, was far from adhering blindly to Kleinism. He rather used it as a tool that he considered particularly appropriate for the problems he had to solve in his clinical work. Some of the analysts of the second generation, though, would be more than ready to raise Kleinism to the stature of a dogma.

In 1951, Pichon Rivière (who was by then president of the APA) and his wife, Arminda Aberastury, also travelled to Europe. In London, Aberastury was supervised by Melanie Klein, with whom she had corresponded since 1945. Both of them had reunions with other renowned Kleinians as well (such as Joan Rivière, Paula Heimann, Herbert Rosenfeld, M.G. Evans and W.C. Scott), to discuss child analysis and the treatment of psychosis.[32] A few weeks later, Melanie Klein would write to Aberastury: 'In my conversations with Betty and Dr Garma I took the same pleasure that I had had with you and your husband. It is very edifying to discuss with people who have such a good understanding of my work'.[33] After London, they went to Paris, as official guests to the *XIV Conférence de Psychanalystes de Langue française*. There, Pichon Rivière met Daniel Lagache, Jacques Lacan, Henry Ey and even Tristan Tzara and André Breton.[34] At the conference, Aberastury presented a paper on transference in child analysis, whereas her husband's communication dealt with transference in psychotic patients.[35]

The following year, Daniel Lagache would publish these two very Kleinian works in the *Revue française de psychanalyse*, of which he had just been appointed editor-in-chief. He would also include his own translation of 'The origins of transference', the paper that Melanie Klein had presented in Amsterdam the year before, in the 17th International Psychoanalytic Congress.[36] That was a considerable amount of Kleinian material for the organ of a society that was not at all Kleinian. Was it maybe the result of a political operation, orchestrated by Lagache, in order to modernize a rather conservative SPP, that he would soon abandon along with Jacques Lacan? In any case, thanks to the links between Lagache and Baranger, there had already been two Kleinian articles written by Garma in the French journal, in addition to another one by Marie Langer, that was to appear soon.[37] Moreover, in 1954, Baranger would translate, with his wife Madeleine, the first book written by a Hispanic analyst to be published in France. It was *La psychanalyse des rêves*, by Ángel Garma, which made part of the 'Bibliothèque de psychanalyse et de psychologie clinique', created by Lagache in the Presses Universitaires de France (PUF).

By 1957, Baranger had already translated a second book by Garma that appeared in the same collection.[38] It was a quite remarkable fact, considering that, at the time, not a single work of Melanie Klein's had been translated into French. This privilege of the Hispanic Kleinian was certainly related to the personal friendship that Garma had with Lagache since the 1930s. But it was probably linked as well to the skills of his personal ambassador: Willy Baranger. If Baranger's activities yielded good fruits, this was particularly true in the direction Buenos Aires–Paris (and even in the vector London–Paris), but not in the sense Paris–Buenos Aires.[39] In those years, the *Revista de Psicoanálisis* published in Argentina did not significantly increase its quota of French articles. At least until 1953, the amount of papers and book reviews by Anglo-American authors continued to be clearly higher than those produced by the French ones, in a proportion close to 10 to 1.

During this period, curiously enough, everything shows that in Paris the works of some Argentine Kleinians were well received, whereas those written by the British Kleinians themselves did not enjoy the same fortune. At the same time, in Buenos Aires, the English and the Americans were more easily read than the French... In short, the Argentine psychoanalytic production – that was mainly Kleinian – was better received in France than Melanie Klein herself. Thus, in 1957, she would write to Lagache to express her disappointment with the French publishing houses:

I am turning again to you for help. Madame Spira has written to me that she will have [the translation of] my '*Contributions* ...' ready for the summer [...]. She has written to Payot, who, in a very polite way, has refused to publish the book. The reason he gives is that he already has a great number of books on this subject. Could you suggest to me another publisher? Baranger is also getting ahead with the translation of the *'Developments...'* I do not want to repeat myself, but you know that I feel very badly about the Presses Universitaires willing to postpone the further publications for years, and I have no time for waiting so long [...]. The trouble is, of course, that my name is not known in France.[40]

It is very striking to note that, by the same time, Ángel Garma, a psychoanalyst working in a peripheral country, had succeeded in publishing two books in the PUF, while Melanie Klein would have to wait until 1959 – a year before her death – in order to see her first book published:

Cher Lagache:

It gave me great pleasure to receive the first copy of *The Psycho-Analysis of Children* from you with your kind inscription, and I thank you very much for the help you have given me in this matter. The book looks quite attractive and I hope it will help to get my work better known in France and also benefit some section of the French population. In any case, it is the fulfilment of a wish, which I have had for the last 26 years. Need I say more?[41]

* * * * *

The first reception of Kleinism in Argentina and France, as we have seen, was rather peculiar. In Argentina, the analytic movement rapidly adopted the ideas of Melanie Klein. These ideas were collectively assimilated, constituting an identity trait. From Argentina, between the 1950s and the 1970s, Kleinism would even be exported to other Latin American countries, such as Uruguay, Brazil and Mexico.[42] In spite of Klein's desires and in spite of geographical proximity, the reception of her theories in France was much more limited and delayed. Kleinian ideas would never have the impact that they had across the ocean, in the southern hemisphere. In this chapter, we have identified some of the contextual elements, the personal biographies and the casual encounters that intervened in this complex process. These factors sometimes combined in paradoxical manners, giving place to unexpected facts and to improbable relations. Thus, in the triangle Paris–London–Buenos

Aires, concerning the 'adventures' of Kleinian psychoanalysis, we have also emphasized how the unlikely became possible and the most likely never occurred.

Concerning the theoretical problems of transmission and reception, this chapter shows the need to question the traditional historical approaches in the field of psychoanalysis, which generally conceive this kind of processes like a unidirectional flow of information, going from the centre to the periphery. Most of the historical narrations in this domain – even some critical ones – have only considered the history of Freudism in "colonial" or "evangelic" terms, as a world-wide dissemination of a series of concepts and practices that were produced in Europe at the beginning of the twentieth century. Even if this model may seem pertinent for a large majority of cases, it tends to ignore or to underestimate any historical process going in the opposite direction, from the periphery to the centre. Therefore, in this chapter, we have tried to highlight the complexity of transmission and reception in a trans-national level, emphasizing the multiple directions that may be involved in this type of operations. As in the history of other disciplines, like psychology, over a hundred and fifty years after the birth of Sigmund Freud, perhaps it is time for the history of psychoanalysis to adopt a more "polycentric attitude," in order to seize the richness of some topics which have not received thus far all the attention that they deserve.[43]

Notes

1. Letter from S. Ferenczi to S. Freud, June 30, 1927. Quoted by P. Grosskurth, *Melanie Klein: her world and her work*, 2nd edn (Cambridge: Harvard University Press, 1987) [1st edn., New York: Knopf, 1986], p. 162. Also available in *The Correspondence of Sigmund Freud and Sándor Ferenczi, 1920–1933*, volume 3 (Cambridge: Harvard University Press, 2000), p. 311.
2. E. Jones, *Papers on Psychoanalysis* (London: Baillière, Tindall & Cox, 1938). See the introduction to his conference "Early female sexuality," quoted by J.-A. Miller, *Le principe d'Horacio* (Paris: Atelier de psychanalyse appliquée, 2001), p. 8.
3. M. Klein, *Die Psychoanalyse des Kindes* (Vienna: Internaler Psa. Verlag, 1932). See also M. Klein, *The Psycho-Analysis of Children* (London: Hogarth, 1932).
4. E. Jones, "Mensajes de cordialidad," *Revista de Psicoanálisis*, 1, no. 1 (1943) 3.
5. See M. Plotkin, *Freud in the Pampas. The emergence and development of a psychoanalytic culture in Argentine* (Stanford: Stanford University Press, 2001), pp. 60–61.
6. See J. Balán, *Cuéntame tu vida: una biografía colectiva del psicoanálisis argentino* (Buenos Aires: Planeta, 1991) and Plotkin, *Freud in the Pampas*.
7. See Plotkin, *Freud in the Pampas*, p. 49.

8. Later on, other Jews with foreign accents would join the group, like Heinrich Racker. Nevertheless, they would always be a minority.

9. For example, in August 1943, Pichon Rivière presented a communication in the APA on the "dynamisms of epilepsy," employing already many Kleinian categories. See E. Pichon Rivière, "Los dinamismos de la epilepsia," *Revista de Psicoanálisis*, 1, no. 3 (1944) 340–381. Reissued in 1983 in *La psiquiatría, una nueva problemática. Del psicoanálisis a la psicología social*, volume 2 (Buenos Aires: Nueva Visión, 1983), pp. 91–134. Langer would do as much the same year, explaining the "psychology of menstruation" in terms of oral sadism and internal fantasies. M. Langer, "Psicología de la menstruación." *Revista de Psicoanálisis*, 2, no. 2 (1944) 211–232. Garma would wait until the following year to praise Klein's ideas on the psychogenesis of maniac-depressive states. See P. Heimann, "A contribution to the problem of sublimation and its relations to the process of internalisation," *International Journal of Psychoanalysis*, 23 (1942) 8–17. This article was reviewed by Á. Garma in *Revista de Psicoanálisis*, 3 no. 1 (1945) 171–172. For a more detailed study of some of the first Kleinian papers in the *Revista de Psicoanálisis*, see H.R. Etchegoyen and S. Zysman, "Melanie Klein in Buenos Aires: Beginnings and developments," *International Journal of Psycho-Analysis*, 86, no. 3 (2005) 869–894.

10. In that period, with the help of his wife, Cárcamo had written the first Spanish translation of Anna Freud's *The Ego and the mechanisms of defence* (London: Hogarth, 1937), published as *El yo y los mecanismos de defensa* (Buenos Aires: Paidós, 1949). See Plotkin, *Freud in the Pampas*, p. 247.

11. See M. Langer, 'El mito del "niño asado"' *Revista de Psicoanálisis*, 7, no. 3 (1950) 389–401 and M. Langer, *Maternidad y sexo* (Buenos Aires: Paidós, 1951).

12. Bela Székely (1892–1955) was a Hungarian psychologist who had studied in Hamburg and Vienna. He practiced psychoanalysis in Budpest, between 1935 and 1937. In 1938, he arrived in Buenos Aires, where he founded the Sigmund Freud Institute. In a very atypical career, he finished by combining his Adlerian ideas with an interest for psychological tests. See G. García, *La entrada del psicoanálisis en la Argentina* (Buenos Aires: Ediciones Altazor, 1978), p. 179. See also G. García, *El psicoanálisis y los debates culturales. Ejemplos argentinos* (Buenos Aires: Paidós, 2005).

13. M. Klein, 'Primeros estadios del conflicto de Edipo y de la formación del superyó', *Revista de Psicoanálisis* 1, no. 1 (1943) 83–110. See also M. Langer, *From Vienna to Managua. Journey of a psychoanalyst* (London: Free Association Books, 1989), p. 195.

14. See A. Aberastury, *Teoría y técnica del análisis de niños* (Buenos Aires: Paidós, 1962). The dates and the information given in this book are not always very accurate.

15. It is very likely that Aberastury read the American version of Anna Freud's *Introduction to the Technique of Child Analysis* (New York: Ayer Company Publishers, 1928). This book would not appear in England until 1946. As to Sophie Morgenstern, the reference was an article of 1939: 'Le symbolisme et la valeur psychanalytique des dessins infantiles', *Revue française de Psychanalyse*, 12, no. 1 (1939) 39–48.

16. A. Aberastury, 'Actualizaciones de la técnica kleiniana en psicoanálisis de niños', *Revista de Psicoanálisis*, 31, no. 5/6 (1983) 21–46.

17. Elizabeth Goode de Garma, interview by the author, Buenos Aires, March 21, 2002 (she died in February 2003).

18. While Freud argued that the psychotic got away from reality because of the Id, Garma affirmed that he yielded masochistically to the sadism of the Superego, which implied a drive renunciation (because of the Id's repression) as well as the abandon of reality. Á. Garma, 'La realidad y el ello en la esquizofrenia', *Archivos de Neurobiología*, 11, no. 6 (1931) 598–616. Re-published in 1932 in *Internationale Zeitschrift für Psychoanalyse*, no. 18, 183–200 (as "Realität und Es in der Schizophrenie") and in 1944 in *Revista de Psicoanálisis*, 2, no. 1, 56–82, with the title 'La realidad exterior y los instintos en la esquizofrenia'.

19. P. Heimann, 'A contribution to the problem of sublimation and its relations to the process of internalisation', *International Journal of Psychoanalysis*, 23 (1942) 8–17. Reviewed by Á. Garma, in *Revista de Psicoanálisis*, 3, no. 1 (1945) 171–172.

20. See Á. Garma, 'Vida y obra de Karl Abraham', *Revista de Psicoanálisis*, 26 (1969) 463–483.

21. Later on, Arnaldo Rascovsky, converted to Kleinism, would push this mandate to the limit, theorising the existence of a 'foetal psyche'. See A. Rascovsky, 'Beyond the oral stage', *International Journal of Psychoanalysis*, 37, no. 4/5 (1956) 286–289 and A. Rascovsky, *El psiquismo fetal* (Buenos Aires: Paidós, 1960).

22. Enrique Pichon Rivière, interviewed by Vicente Zito Lema, Buenos Aires, 1976. See V. Zito Lema, *Conversaciones con Enrique Pichon Rivière sobre el arte y la locura*, 13th edn (Buenos Aires: Ediciones Cinco, 1980) [1st edn: Buenos Aires: Timmerman Editores, 1976], p. 49.

23. Couchée dans une forêt/verte et vierge/la femme noire dans ce temps/serre son cœur palpitant/entre ses jambes/rondes et charnues/attendant/le visiteur égaré/sortir de l'épaisseur/de sa nuit fantasmale [sic]. Zito Lema, *Conversaciones con Enrique Pichon Rivière*, pp. 32–33.

24. Pichon Rivière, *Revista de Psicoanálisis*. Reissued in Pichon Rivière, *La psiquiatría*, pp. 91–143. Communication presented in the APA on August 26, 1943. Later on, in his theory of 'the unique illness', the paradigmatic case would shift to melancholy. However, the psychopathological mechanism would stay unchanged: frustration, ambivalence, aggression turned against the Ego, introjection.

25. See H. Vezzetti, *Aventuras de Freud en el país de los argentinos* (Buenos Aires: Paidós, 1996), pp. 253–258.

26. See A. Dagfal, 'El pensamiento francés en la Argentina: el caso de los "discursos psi"' *Conceptual, estudios de psicoanálisis*, 6, no. 7 (2006) 11–16.

27. See S. Resnik, 'Préface' in W. Baranger (ed.), *Position et objet dans l'œuvre de Melanie Klein* (Paris: Erès, 1999), pp. 11–16.

28. Letter from Willy Baranger to Daniel Lagache, October 14, 1947 (we thank Éva Rosenblum, in charge of Daniel Lagache's correspondence, in Paris, for giving us access to this letter).

29. See É. Roudinesco, *Jacques Lacan & Co.: a history of psychoanalysis in France, 1925–1985* (Chicago: University of Chicago Press, 1990).

30. However, according to Elizabeth Goode de Garma (better known as 'Betty'), her husband had already met M. Klein, although she could not precise the date nor the circumstances of their first encounter. Elizabeth Goode de Garma, interview by the author, Buenos Aires, 21 March 2002.

31. Elizabeth Goode de Garma, interview by Emilia Cueto, Buenos Aires, date unknown, http://elsigma.com/secciones/entrev.jsp?id_entrev=795 (accessed October 25, 2004).

32. Notas e informaciones, *Revista de Psicoanálisis* 8, no. 4 (1951) 594–595.

33. Letter from Melanie Klein to Arminda Aberastury, February 7, 1952. Quoted by S. Fendrik, *Desventuras del psicoanálisis* (Buenos Aires: Espasa Calpe, 1993), pp. 48–49.

34. Zito Lema, *Conversaciones con Enrique Pichon-Rivière*, pp. 55–56.

35. A. Aberastury, 'Quelques observations sur le transfert et le contre-transfert dans la psychanalyse d'enfants', *Revue française de psychanalyse*, 16, no. 1/2 (1952) 230–253; E. Pichon Rivière, 'Quelques observations sur le transfert chez des patients psychotiques', *Revue française de psychanalyse*, 16, no. 1/2 (1952) 254–262.

36. M. Klein, 'Les origines du transfert', *Revue française de psychanalyse*, 16, no. 1/2 (1952) 204–214. The same year, it was published in English as M. Klein, 'The Origins of Transference', *International Journal of Psychoanalysis*, 33 (1952) 433–438.

37. See Á. Garma, 'Origine et symbolisme des vêtements', *Revue française de Psychanalyse*, 14, no. 1 (1950) 60–81; Á. Garma, J.C. Bisi and A. Felgueras, 'Les agressions du surmoi maternel et la régression orale-digestive dans la genèse de l'ulcère gastro-duodénale', *Revue française de Psychanalyse*, 15, no. 4 (1951) 527–557; M. Langer, 'Le mythe de l'enfant rôti', *Revue française de Psychanalyse*, 16 (1952) 509–517.

38. It was *La psychanalyse et les ulcères gastro-duodénaux* (Paris: PUF, 1957). Finally, in 1962, a third book by Garma would be published in this collection: *Les maux de tête* (Paris: PUF, 1962). But the translation would be made by Elza Ribeiro Hawelka.

39. To complete the triangle, Baranger would later translate (from English to French) the book *Developments in psycho-analysis*, written in 1952 by Melanie Klein and her closer collaborators. It would be published in 1966, in the collection still directed by Lagache.

40. Letter from Melanie Klein to Daniel Lagache, December 19, 1957 (we thank Éva Rosenblum, in charge of Daniel Lagache's correspondence, in Paris, for giving us access to this letter).

41. Letter from Melanie Klein to Daniel Lagache, March 28, 1959 (we thank Éva Rosenblum, in charge of Daniel Lagache's correspondence, in Paris, for giving us access to this letter).

42. For a more detailed treatment of the spread of Kleinism in Brazil, see our doctoral dissertation, *Entre Paris et Buenos Aires: la construction des discours psychologiques en Argentine (1942–1966)*, defended and passed on June 16, 2005, at the University of Paris VII, under the direction of Elisabeth Roudinesco.

43. See K. Danziger, 'Towards a polycentric history of psychology', conference given in 1996, in the 26th International Congress of Psychology, held in Montreal.

8
The Lacanian Movement in Argentina and Brazil: The Periphery Becomes the Center

Jane Russo

In the turn of the century, it became commonplace to speak of the decline or of the imminent disappearance of psychoanalysis. In face of the fast progress of neuroscience and the successful pharmacological influence on psychiatry,[1] psychoanalytic vocabulary and logic seemed decidedly outdated. However, we should ask *which point of view* has produced such cliché. Is it possible to mention a decline of psychoanalysis in all the countries where it has laid roots? Or would we be mistaking, as usual, a specific occurrence in one of the central countries for a truly global event? The question is pertinent since the United States are undoubtedly a propagating center for habits and customs, as well as for ideology. Therefore it is common to attribute to the North American hegemonic pattern in several scientific and research areas a worldwide reach that it often does not have (yet).

Despite the fact that in the United States psychoanalysis seems to have lost the battle against the biological trend in psychiatry, I don't think we can generalize this fact to the rest of the western world, particularly in the case of Latin America countries. In those countries, especially Argentina, Brazil and Mexico, the Lacanian trend in psychoanalysis shows renewed energy, expressed in the number of training institutions continuously emerging, in the disagreements and internal disputes within the field, as well as in the congresses and meetings always held with an expressive attendance. At first we notice an alliance between France as a propagating center and Latin America as a receiving center, the traditional role of the periphery in relation to the center. It is as if there has only been a shift in the central country, from England (which dominated the formation of IPA affiliated associations) to France. This, however, is only partially true.

The diffusion of psychoanalysis in Argentina and Brazil are part of a much more complex transnational phenomenon. As I try to demonstrate in this work, in the case of the psychoanalytic movement in Argentina and Brazil and their present developments, a specifically Latin-American trend preceded and prepared the terrain for a "Lacanian revolution." That trend, originated from challenging the monopoly of the IPA-affiliated associations,[2] was fueled by the leftist critical thought which was then affirming itself in the continent, and, starting with Argentine "dissidents," spread in Brazil bringing with it new ways of thinking and practicing psychoanalysis.

The next step consisted of an intense diffusion of Lacanian theory and practice, with the appearance of many groups and training institutions, apparently reinstating the two countries in their traditional role of a receiving periphery. However, the Lacanian rhetoric, in an attempt to retrieve the charismatic force of the movement's origins, encouraged a certain degree of decentralization and autonomy both in the transmission of ideas and in the theoretical production, which led to an endless proliferation of ruptures, disputes and consequent appearance of new associations, in fact causing a certain degree of inversion (or, at least, a relative symmetry) in the center-periphery relationship.

This is the history that I attempt to recover in this work, focusing specifically Brazil and Argentina.

In order to do so, I begin with a general discussion on the meaning of the Lacanian "revolution" in psychoanalysis, trying to interpret it from the standpoint of the opposition between bureaucracy and charisma inspired in Max Weber's theory.[3] As a next step, I try to present a brief history of the psychoanalytic movement in Argentina and Brazil, detailing the decades that preceded the arrival of Lacanism in each country, and also the influence exercised by Argentine psychoanalysis over the Brazilian psychoanalytic movement. Next I try to associate the reception of Lacan's ideas to the tensions that permeated the "psy" field in each country, showing that there was a common ground for such tensions. I finish with a few reflections about the current situation of Lacanism in Brazil.

The reader will notice that this is a work written by a Brazilian, who conducted research on the psychoanalytic movement in Brazil. In other words, the material on the Brazilian movement is more extensive and detailed than the material on Argentina,[4] therefore causing a certain degree of imbalance.[5] Nevertheless, I hope I am able to demonstrate that the Brazilian psychoanalytic movement cannot be understood without taking into consideration the Argentine movement.

The Lacanian revolution

The diffusion of psychoanalysis as a cultural phenomenon took place in parallel to the appearance of training associations and intense disputes, within the professional field thus formed, over the monopoly and transmission of the title of psychoanalyst title. Such disputes and disagreements were fueled – both at the level of theory and at the political level – by the tension, inherent to the field itself, between bureaucracy and charisma.

A clear example of such tension was the double power-mechanism that became effective within the psychoanalytic movement during the second decade of last century. The International Psychoanalytic Association was founded in 1910, with Freud's support. In 1912, soon after IPA's creation, Freud instituted a "secret committee," whose members were, besides himself, five early disciples (a sixth one was added in 1919), all of them Jews, except for Ernest Jones. Up to 1927 – when the secret committee was dissolved – the two institutions worked in parallel: the IPA and the Committee.[6] The former developed as a bureaucratic association that tended to establish regulations and impersonal statutes, while the latter operated as a sort of "secret society," organized around the charisma of the founding father. This is why psychoanalysis, from its early days, has been a *sui generis* science, lacerated between the "respectability" of the sanctioned sciences and the marginality of a kind of "occult" knowledge depending on some form of initiation.

Such tension generated a certain amount of dualities, among which two are worth mentioning: one geographical and the other professional. It is a well-known fact that during the 1930s, with the rise of the Nazi regime in Germany and the subsequent annexation of Austria, German psychoanalysis was practically extinguished. Until then a propagation center for psychoanalytic doctrine and practice – particularly through the leadership exercised by the Berlin Institute – Germany was replaced by Anglo-Saxon countries – England and the USA. The English language replaced German as the language in which the psychoanalytic teaching was transmitted. Such replacement had the significant consequence of bringing psychoanalysis close to medicine, in both countries.[7]

We then arrive at the second duality that apparently characterizes the whole psychoanalytic movement: the tension between medical doctors and "laymen," clearly subordinated, as we will see, to the tension already pointed out between bureaucracy and charisma (or the

bureaucratic versus the charismatic form of control). The rapprochement to medicine favors the "officialization" of psychoanalysis: control
of the monopoly and of the transmission of the title through clear training rules. The "lay" training, on the contrary, may (although it does not
necessarily do so) approximate psychoanalysis to the humanities and to
artistic avant-garde movements that, as we know, took an early interest in the doctrine. On the one hand we have a kind of "subversive"
knowledge that causes resistance because it opposes established morals
and routine beliefs and, on the other hand, we have the respectability of a science that contributes to the good order and maintenance
of society.

The well-known English and American pragmatism fed one of the
sides of such duality, emphasizing the search for practical results in the
psychoanalytic clinical work.[8] Thus a "psychology of the ego" emerged
that, as the name indicates, emphasized the adaptive and regenera-
tive properties of the ego. Another trend also emerged in the US: the
"culturalist" school that, maintaining a close dialogue with anthropology and other social sciences, severed ties with IPA and founded the
International Federation of Psychoanalytical Associations (IFPS) in 1962.

With the exception of the so-called culturalist dissidence, all doc-
trinaire disagreements remained inside the psychoanalytic movement,
absorbed by IPA. Even the culturalist advocates, although questioning
IPA's monopoly, did not question the legitimacy of the *way* the analyst's
training should develop. In short, the institutions created under the
aegis of IFPS followed the IPA model, which continued with the patterns
imposed by the former Berlin Institute. In that sense the Lacanian rupture was radical, taking to the extreme the tension between bureaucracy
and charisma.

Lacan's motto, as we know, was the "return to Freud," who, according to him, was being forgotten by the post-Freudians leading the then
existing associations. In a sort of fundamentalist movement, Lacan proposed to restore the heroic times of the founding father and to recover
the subversive face of psychoanalysis. This was neither a mere theoretical disagreement nor a dispute over the granting of the psychoanalyst
title to "laymen."[9] This was a denounce of "deviations" from "true"
psychoanalysis.

The famous aphorism "the analyst authorizes himself" is a war-
cry against formal and bureaucratic rules, favoring a return to the
"charismatic transmission" of the psychoanalytic knowledge. It is during the singular and unique process of analysis that a person becomes
an analyst. Something magical or intangible is restored with such

formulation. At the same time the psychoanalytic training (currently called "transmission")[10] becomes endless, since there is not (at least in the beginning of the "Lacanian revolution") a credential that is granted once and for all. It is expected from the one who calls him/herself a psychoanalyst a visceral engagement to the psychoanalytic theory and practice as well as to his/her institution. S/he must constantly – through participation in "cartels," written works, clinical sessions etc. – "give testimony" of unrestricted adhesion to the cause.

The famous variable-time sessions, which usually do not exceed 20 minutes, provide a good measure of what it means to adopt a charismatic type of control versus a bureaucratic one. Justified by theory, inasmuch as the unconscious temporality (the so-called "logical time") cannot be regulated by the clock, the variable-duration sessions imply the impossibility of determining in advance the singular and random moment when the unconscious breaks through. It is the analyst's task to notice and acknowledge such a moment, interrupting the session. Evidently the way of noticing it or interrupting the session does not follow any predetermined formula, and what we could call the analyst's "talent" (part of his or her charisma) is what determines the course of events. Bureaucratic control presupposes, on the contrary, a 45 or 50 minutes session with a clearly delimited beginning and end, seeking a rational organization of the analyst's and the patient's time. In the case of the variable duration session, the analyst's and the patient's time are disorganized from the standpoint of bureaucratic rationality, obeying a charisma that belongs, at the same time, to the analyst (responsible for noticing and capturing what is ineffable), and to a situation which is itself unusual (a subject that speaks whatever comes to his mind while lying with his back to a stranger).

Similarly to what happens in the transmission of knowledge and in the session format, form and content are also inseparable concerning theoretical production. When one discusses psychoanalysis, or writes about it, one cannot use the same language as in traditional sciences, the language of consciousness and reason. To understand and discuss psychoanalysis we need other forms of reasoning and language, which incorporate the way the unconscious operates. Those who have read, or tried to read, Lacan's texts, know what is meant here. There is absolutely no concession to pedagogy. The wordplays, the double meanings, ellipses, metaphors, and other figures of speech are the tonic of his texts. The *form of the text, how it is written* is as important as its content. Such submission to unconscious rules of production makes the

Lacanian writing extremely sophisticated – approaching a sort of literary production – and his reasoning extremely difficult to understand by non-initiates.

The absence of any "pedagogic" attitude, both in the transmission of theory and in the clinical work, is related to the criticism of the "adaptiveness" and the pragmatism of North American psychoanalysis. Lacan produced an eminently French psychoanalysis. Following the linguistic structuralism of the sixties and totally immersed in the intellectual context of that time, Lacan's theory became a sophisticated consumption-good for a thinking elite, eagerly consumed by intellectuals, artists, writers, movies critics, thus moving away from psychiatrists and other health professionals. The issue, however, was not merely to move away from medicine's most pragmatic concerns, but also to build a barrier against the intense diffusion, or even the banalizing of psychoanalysis, as seen on the other side of the Atlantic.

Concerning the monopoly and control over the granting of the coveted psychoanalytic credentials, however, saying that "the analyst authorizes himself" does not mean that anyone can become a psychoanalyst. On the contrary, instead of requiring a specific degree, as was the case in North American associations [11] what is required by Lacan's followers is the capacity to apprehend and reproduce a highly formal and abstract speech (and also to recognize it as a cultural asset holding a high symbolic value). At the same time, by refusing easy formulas or didacticism, Lacan's writing intended to be a barrier against the social banalization of psychoanalysis.

* * * * *

The attempt by the "Lacanian revolution" to reconstruct a charismatic control of the transmission of psychoanalytic credentials, in detriment of bureaucratic control, did not take place without contradictions or conflicts. These were present in the movement since its inception, and apparently have become stronger.

After Lacan's death, with the ascension of his son-in-law Jacques Alain Miller as guardian of the master's inheritance, things have become even more complex. The foundation of a World Association of Psychoanalysis (also known as the Lacanian IPA), orchestrated by Miller, illustrated the seemingly insoluble tension between a purely charismatic operation and the need for bureaucratic controls in order to maintain the monopoly exercised by the training institutions.

The appearance and diffusion of psychoanalysis in Argentina

The diffusion of psychoanalysis – as a practice and as a world view – both in Brazil and in Argentina has been a well-known phenomenon within the international psychoanalytic community, the social sciences and psychoanalysis scholars.

Studies indicate that by the time when the "official" training institutions were founded, in the forties and fifties, there had been already in the two countries a significant interest for the new therapy and even an incipient consumer public.[12]

The Asociación Psicoanalítica Argentina (APA) (Argentine Psychoanalytic Association) was founded in Buenos Aires in 1942 and recognized officially by IPA in 1947. Unlike Brazilian institutions, it did not depend on the "import" of European psychoanalysts, having an eminently "local" beginning.[13] The extreme centralization of Argentine cultural life around Buenos Aires made APA the only "official" psychoanalysis training institution in the country until 1976.[14] Although having admitted non-doctors at the beginning of its operations, in the fifties APA restricted admittance to candidates with a medical degree. With the appearance of psychology undergraduate courses in the late fifties, all of them extremely permeated by psychoanalytic theory, an unsustainable situation began to emerge, similar to what also happened in Brazil. The excessive demand for psychoanalysis training found an insurmountable barrier in the scarce offer represented by APA. Psychologists were refused access to *legitimate* psychoanalytic training. At the same time, a series of "unofficial" supervision and study groups coordinated by APA-trained psychoanalysts offered non-sanctioned psychoanalytic training to people without an MD.

The intense political polarization that Argentine society suffered during the 1960s had significant consequences in the psychoanalytic field. In 1969, within APA, the "Platform" group was formed by leftist psychoanalysts who demanded from the association a political positioning concerning what was happening in Argentine society. The criticism to an "alienated" psychoanalysis, practiced in private clinics, disconnected from the surrounding reality, found an echo in another group of psychoanalysts know as the "Document" group. The two groups – Platform and Document – left APA in 1971 and joined the Argentine Federation of Psychiatry (FAP) and the Buenos Aires Psychologists Association (APBA), creating the Coordination Bureau for Mental Health Workers, with the purpose of acting in psychiatric hospitals, unions and low-income communities. Psychoanalysis remained the reference for the work being developed. To the large number of psychologists graduating at the time, the Coordination Bureau represented a somewhat unorthodox opportunity to work in psychoanalysis.

The struggle for psychoanalysis social engagement led progressive psychoanalysts to attempt an articulation between the Freudian and the Marxist theories. Actually, according to Plotkin, psychoanalysis became in that period an important element of the leftist culture, and this proved crucial to the wider acceptance of psychoanalysis in the country. It is at that moment that the Lacanian thought emerged and started to consolidate in the Argentine psychoanalytic scenario. Even if the joining of psychoanalysis and leftist thought was not necessarily questioned, Lacanian speech proposed another way of establishing an articulation between psychoanalysis and revolution, by emphasizing the intrinsically revolutionary nature of psychoanalysis as a science. The revolutionary character of psychoanalysis did not, therefore, originate in its articulation to Marxism, or in the type of social engagement exercised by psychoanalysts, but in its being a science of the unconscious and of desire.[15]

Lacanism, with its anti-APA rhetoric (which at that time represented the establishment and the oppression that must be opposed), its alliance with the artistic and literary *avant-garde* movements, and the proposed recovery of primitive psychoanalysis virulence, accentuated the ties between leftist thought and psychoanalytic theory.

From the psychologists' point of view, the new theory meant the possibility of bringing together, on the one hand, political extremism and, on the other, the fight against the monopoly of psychoanalysis training held until then by APA.

The first Lacanian group of Argentina emerged in 1969, the Lacanian Group of Buenos Aires, composed by leftist intellectuals and revolving around Oscar Masotta. This group was the embryo of the first Lacanian institution of the country, the Freudian School of Buenos Aires, founded in 1974. Masotta, a legend in the Argentine intellectual milieu, was a self-taught intellectual oriented to philosophy and literature, as well as to *avant-garde* artistic movements. His influence over the intellectuality of the period was immense. After the establishment of the Freudian School in Buenos Aires, he went to Spain, where he worked for the diffusion of the Lacanian theory until his death in 1979.[16] As was the case with the APA, with Masotta and his Freudian School we have again a "local" appearance of Lacan's thought in Argentina.

In the eighties there was an intense blossoming of Lacanian training societies, similar to what happened in Brazil. We could say that, repeating what seems to be an inevitable destiny, the theory that was to be "marginal" turned out to be hegemonic.

Brazil: stability in São Paulo and fragmentation in Rio de Janeiro

The emergence of psychoanalysis in Brazil was somewhat more complex than in Argentina. First because during the first decades of the twentieth century, Rio de Janeiro, then the capital of the country, shared with São Paulo cultural and economical hegemony over the country. In this sense, there was a double emergence of psychoanalysis: in São Paulo and in Rio de Janeiro. In spite of an attempt made by the paulista group to achieve autonomy,[17] in both cities the training institutions originated from "emissaries" sent by the IPA in response to the local physicians' demand.

In December, 1936, Adelheid Koch arrived in Sao Paulo, a Jewish doctor recently graduated from the former Berlin Institute of Psychoanalysis. She became the first "training analyst" in Brazil, in charge of organizing a study group sanctioned by the International Psychoanalytic Association. The group was officially recognized by the IPA in 1951, as the Brazilian Society of Psychoanalysis of São Paulo.

In Rio de Janeiro, where the psychiatric milieu was already strongly structured, psychoanalysis, although circulating among illustrious names of the profession, took a long time to become an interesting alternative to psychiatric practice *stricto sensu*. A group of young psychiatrists, who usually met to study and discuss psychoanalysis, feeling tired of waiting for a movement from the IPA, decided to go to Buenos Aires for training. Meanwhile, in 1948, the IPA sent two psychoanalysts to begin the training of analysts – Mark Burke, member of the British Psychoanalytic Society, and Werner Kemper, member of the Berlin Society. Two years later, divergences among the two analysts led to the formation of two groups, one guided by Kemper and the other by Burke. The psychoanalysts who had graduated in Buenos Aires joined Burke's group. The German analyst's group ended up being recognized by the IPA in 1955 as the Rio de Janeiro Psychoanalytic Association – SPRJ, and the other group, made up of former Burke-analyzed patients and by the Argentine-trained analysts, was recognized in 1957 as the Brazilian Society of Psychoanalysis of Rio de Janeiro – SBPRJ. Unlike the São Paulo association, which, consistent with early established bonds with the non-medical milieu, admitted "lay" candidates, the two Rio associations were quickly monopolized by psychiatrists and, very much like the American associations, soon started to require a medical degree from candidates. As we see, the warm reception of psychoanalysis on the part of Rio's medical establishment was not without consequences, and a privileged bond with psychiatry was the major characteristic of the movement in Rio de Janeiro.

The opening to non-medical candidates granted the São Paulo movement and the São Paulo society in general a high degree of stability and relative peacefulness, when compared to the similar Rio institutions.[18] In Rio, as we saw, there was immediate dissidence, causing the proliferation of associations fueled by the closing of the "official" associations to non-doctors. In the late seventies the city was certainly a record breaker, having at the time nearly 20 psychoanalytic associations of varied shades.[19]

Therefore, when comparing Brazil and Argentina, we see the key role and total monopoly of APA in the Argentine case, in contrast with the Brazilian decentralization and the Rio de Janeiro movement fragmentation. We already saw that Rio psychoanalysis, at origin, had a narrow bond with the APA, through APA – trained analysts who became founding members of the SBPRJ, teaching at the institution and becoming later prominent analysts. The influence of Argentine psychoanalysis over Brazilian psychoanalysis intensified during the seventies, a moment when the two countries similarly underwent an unprecedented psychoanalytic boom, in addition to strong dictatorial regimes.[20]

Sixties and seventies in Brazil: psychoanalysis without politics?

The military coup of 1966 in Argentina deepened the process of political radicalization that society was living since the Perón administration (1946–1955). The "psy" professionals were involved in political engagement and psychoanalysis was a part of the leftist thought of the early seventies. In fact, the dictatorship established in 1976 seems to have affected many of those professionals, whether with imprisonment, torture and death, or with banishment or exile (often voluntary).

It is usual to compare the Argentine and the Brazilian cases, arguing that in Brazil the psychoanalytic boom was the result of the de-politicization caused by the military dictatorship – since there was no possibility of acting in the public space, people, particularly young ones, turned to the private sphere in order to face their conflicts and contradictions. In fact, the "official" psychoanalytic associations did not speak out against the repression conducted by the Brazilian state.[21] However, one should reformulate the widely spread idea of an alienated "AI-5 Generation"[22] seeking refuge in drugs and on the couch.[23] One cannot deny the almost total impossibility of political engagement using the traditional ways. Another form of engagement, however, that could be called "countercultural," attracted young people involved in the culture of drugs, poetry and "alternative" art in general.[24] Resistance spaces were created and valued by the post-AI5 generation, leading to the emergence of themes considered smaller by the traditional left, such as sexual and gender behavior, and bringing to the fore the so called "minorities" (blacks, homosexuals, women, insane people, indigenous populations, etc.) that tended to take the place once occupied by the working class.

That proliferation of an opposing "alternative" culture took place simultaneously with the "authoritarian modernization" produced by the military regime, affecting particularly the media, and causing a significant increase in the possibilities of social ascension by the urban middle class. It is possible to see a close relationship between social mobility and psychologism. Upward social mobility cannot be understood merely from an economic point of view, since it usually implies change of address, change in consumption habits and, often, changes in behavior, above all in the younger generation. All this tends to produce a certain degree of disorientation, as the values from the original background must be forsaken (or denied) on behalf of the new lifestyle. Denial of the ascending trajectory is usually part of such abandonment, creating the need to attribute the resulting discomfort or disorientation to inner difficulties, belonging to the psychological realm.[25]

In this sense, the universe of countercultural criticism that challenged traditional behaviors concerning both sexuality and family values as well as labour life was not completely foreign to the crisis of values caused by the social ascension of middle-class sections that, somehow, also put in check traditional behaviors related to the family's original background. In both cases the possible result tended to be a search for self-identity, for true desires and aspirations (whether through therapy, drugs, or, quite commonly, both). It is possible to say, therefore, that psychologism (and the psychoanalytic boom) has a shared paternity. At the same time that it seems to be a child of political repression, it is also a child of countercultural protest and of the social mobility of the middle strata of the urban population.

Psychoanalysis boomed impressively during the so-called "years of lead" in Argentina and Brazil. We saw that in the Argentine case, such boom was connected, at first, to the blooming of a vigorous leftist thought, which is not the case in Brazil. However, a phenomenon approximated the two countries, with similar consequences in both places. A new actor entered the scene, the psychologist, an avid consumer of psychoanalytic theory and therapy, but with his or her access to the psychoanalyst title prevented by official associations[26].

The Argentine "invasion"

In the two countries the blooming of university programs on psychology were completely harnessed to the diffusion of psychoanalysis beyond the medical profession.[27] And, as in the Argentine capital, Rio de Janeiro psychologists, themselves psychoanalysis patients supervised by psychoanalysts from the official associations, could not officially practice psychoanalysis. A change in such situation began during the seventies related to a sort of "Argentine invasion" that preceded (and possibly prepared) the "French invasion" represented by Lacanism.

According to Ana Cristina Figueiredo, the Argentines role in the organization of the psychologists' movement was crucial. They arrived mostly through APPIA, Association of Psychiatry and Psychology for Childhood and Adolescence, founded in January, 1972, in Rio de Janeiro. APPIA followed the model of ASAPPIA (Argentine Association of Psychiatry and Psychology for Childhood and Adolescence), whose purpose was to promote debate among professionals of the three areas – psychiatry, psychology and psychoanalysis – at an international level through congresses, publications, and lectures. This was the group that sent the first Argentines to Rio in order to teach courses at SPRJ.[28] The invitation was made by Fábio Leite Lobo (then director of the SPRJ Teaching Institute) and it finally provoked a crisis that resulted in the creation of the first graduate institute of psychoanalysis devoted exclusively to psychologists.[29]

The IOP, founded by Fábio Lobo and a group of psychologists, and especially the APPIA propagated the "Argentine style" of psychoanalysis,[30] and attracted an enormous contingent of psychologists around its project. Commenting on IOP's courses contents, Ana Cristina Figueiredo emphasizes the importance of the "Argentine style" for the statement of professional autonomy on the part of Brazilian psychologists:

[The Argentine psychoanalysts] brought significant innovations to techniques, such as the brief therapies employed in institutional works, as well as group techniques. Furthermore, the very style of working has its peculiarities: the rupture of certain formalities in therapeutic sessions, the number of weekly sessions, the discussion on neutrality, new approaches for children and adolescents [...] It was even mentioned a type of 'Argentine neo-Kleinianism', whose major theorists were Bleger, Aberastury, Pichon-Riviere, Rodrigué and Rascovsky. One of the consequences was the idea that changing the techniques and working the setting with more flexibility did not question the authenticity of the psychoanalytic model. To the contrary, it promoted and provided input for a review of the Brazilian analysts' orthodoxy. It was exactly what psychologists needed: a legitimate criticism to their own exclusion from psychoanalysis.[31]

According to Manoel Berlinck, the "Argentine invasion" dramatically changed the psychoanalytic scenario in São Paulo as well. He states that the arrival of Argentine psychoanalysts in São Paulo meant breaking of the monopoly held until then by the Brazilian Society of Psychoanalysis of São Paulo, in addition to introducing (as in Rio de Janeiro) a "new psychoanalysis." Among the theoretical-technical contributions brought by the Argentines he mentions, besides Oscar Masotta's interpretation of Lacan, "a concern with the sociopolitical dimensions of psychoanalytic practice, relatively unknown until then in São Paulo."[32] Berlinck has an extremely pertinent comment on this: "Seen in retrospective, APA's scission in 1971, which strongly reflected in São Paulo and Brazil, could be seen as a symptom of the expansion of psychoanalysis in Argentina."[33]

In other words, already "colonized" by the English,[34] Brazilian psychoanalysis underwent a period of "liberation" from Klein's dogma, through a kind of Argentine "colonization," previously to being conquered by the French master.

The French invasion

In fact, the Argentine "invasion" seems to have been the first step in the break of the monopoly exercised by the IPA-affiliated associations. Somehow, as we saw, the Brazilian urban centers had a situation similar to Buenos Aires: a broad diffusion of psychoanalysis, including the "psy" professional milieu, with a restrictive control of training. Similarly, the social diffusion of psychoanalysis – in the broad culture, in the media and among the learned middle segments – was similar in both countries. The Argentine intellectual field, however, differed from the Brazilian in two significant aspects. On the one hand, there were the effervescence and the political extremism that marked the sixties and seventies. On the other hand, there was a more consistent middle class (numerically speaking), more educated than the Brazilian. Those two factors seem to have allowed for a higher level of psychoanalytic vitality in terms of theoretical production and clinical innovation, as compared to Brazil. In this sense, we could metaphorically speak of a "psy" Argentine imperialism in relation to Brazil.

If the innovations brought by Argentine psychoanalysts were the first step in the rupture of "official" orthodoxy, such rupture was only consolidated with the arrival of the Lacanian movement. The movement supplied the theoretical foundation and therefore the required legitimacy for the increasing number of psychologists that wished to practice psychoanalysis and obtain psychoanalytic credentials, as well as young psychiatrists dissatisfied with the forms of recruitment and training of existent associations. The Lacanian path did not merely supply new clinical approaches or even new readings of Freudian texts. More than that, it presented a new logic, an entirely new way of thinking about training emanating from the core of the theory. Thus we return to the argument developed at the beginning of this article. The transformation provoked by Lacanism did not follow the logic of bureaucratic control, *which was external, to a certain point, to the theory*, as it would be the case of changing the associates' participation, or the presence or absence of "training analysts," or the possibility of choosing one's supervisor. Such changes could take place in a way that is reasonably independent from the theoretical option of the group. From the standpoint of Lacanian logic, which is based on charismatic control, as we saw, the changes are part of the theory and only justified if originating there.

In order to illustrate the difference between both types of control, and how such difference appears in the way associations present themselves, we mention here a few excepts taken from the *Agenda of Psychoanalysis*.[35]

The Freudian Letter School (Escola da Letra Freudiana) is a place that privileges Freud's and Lacan's thought in psychoanalysis transmission. Transmission in psychoanalysis complies with the following condition of the Freudian word – not to nourish itself from any knowledge presumed complete, nor to find safe warranties – therefore it is always the transmission of an absence, a place for the inscription of a certain ethics (...) Thus, having the cartels as a base structure, the School defines itself as a tool for permanently forming analysts, through textual transmission and teaching, as well as commitment to written production and clinical practice, sustained by a strict questioning of the direction of the Cure and the Completion of the Analysis.[36]

In order to persevere, to endure and to remain in accordance with the sentence that founds the Freudian Thing – and no practice other than psychoanalysis is directed toward that which is the core of the Real in the heart of experience – without Another, thus devoid of a guarantee. We then instituted a warranty work and to this end, hereby, we institute the degree of Member of the Thing.[37]

Let us compare those true professions of faith to the prosaic SPRJ presentation:

The association has a graduation institute of psychoanalysts. To enroll at the institute the person must be a physician or a psychologist, and enroll at SPRJ itself. There will be three interviews with members appointed by SPRJ and, if the candidate is approved, he or she must begin his or her personal analysis with one of the institution's didactic psychoanalysts, having the right to chose among them.[38]

The difference in the style of language employed is in fact remarkable, expressed not only in the contrast of the terms used but above all in the need that Lacanians have to take a stand in their texts, which need a certain degree of initiation to be understood. Compared to the more or less impersonal and neutral language (based on bureaucratic control) of the "official" association – with terms such as "institution," "course," "candidate," "selection" – the text of the Lacanian associations presupposes not only estrangement from a bureaucratic rationality, but above all a reasonable degree of separation: between those who understand (or at least imagine they do) and those who do not understand; between the initiated and the non-initiated. In a time when psychoanalysis is vulgarized as a cultural artifact, this is not dissemination in the usual sense of the term but, on the contrary, obstruction, containment, delimitation. If an association such as SPRJ tries to provide some sort of delimitation by stating in its text that its graduation course is strictly for holders of a medical or psychology degree, this restriction soon seems naive and even ineffective when we read something like: "'The School is open to all those who have a desire for the teachings and the legacy of Freud and Lacan . . . ". The Lacanian style does not seek popularization in the usual sense, because, requiring some degree of previous initiation, it necessarily implies exclusion. What seems to happen is a popularization through fascination – a fascination exercised both by the elaborated language, sometimes of difficult understanding, and by the borrowed charisma of the emblematic figures of Freud and Lacan.[39] A language that is additionally a source of prestige, because it brings those using it (the initiates) closer to the intellectual vanguard. On the other hand, the use of such language has a significant internal function: the Lacanians and their associations write to be read by their peers (other Lacanians and other associations), in a mutual control of loyalty to the Lacanian word. From this standpoint, the great public remains in the background.

The Lacanians training schools offered, therefore, not merely a strong theoretical-conceptual base for an analytical formation detached from the IPA, but also a demand for strictness and commitment to "the cause" that could not be considered slightly. Responding to the demand by psychologists and other professionals (such as social workers or even social scientists) that wished to become psychoanalysts, the Lacanian associations at the same time made it clear, by their own mode of operation, that such title was not meant for anyone.

The first institutions

Both in São Paulo and in Rio the first Lacanian institutions appeared in 1975. In São Paulo, Luiz Carlos Nogueira, Jacques Laberge and Durval Cecchinato founded on that same year the Center for Freudian Studies. Regional chapters were gradually created (in Recife, Brasília, Salvador, Curitiba, Natal and Porto Alegre). Three years after foundation, CEF faced dissidence, which resulted in the Freudian School of São Paulo, which, on its turn, ceased activities in 1980. In 1982, the Brazilian Freudian Library emerged, organized by Jorge Forbes.[40]

On the São Paulo movement, Berlinck remarks in his 1989 text:

> There is today in São Paulo one single prominent Lacanian psychoanalytic organization, the Brazilian Freudian Library, devoted to transmission. Additionally, there is the São Paulo School of Psychoanalysis, quite recent, and a few groups of restricted significance, such as, for instance, Che Vuoi? In other words, unlike Bahia, Rio de Janeiro and Rio Grande do Sul, where Lacanian psychoanalysis is widespread, in São Paulo it is somewhat limited. Therefore, to this day, and with rare exceptions, traditional intellectuals ignore Lacan.[41]

We can interpret the weaker presence of the Lacanian movement in São Paulo as due to the high stability of the Brazilian Society of Psychoanalysis of São Paulo and to the fact that it admitted non-doctors since its inception, therefore not creating such a strong demand for alternative training.

In Rio de Janeiro Betty Milan e M.D. Magno founded in 1975[42] the Freudian School of Rio de Janeiro (CFRJ). In the late seventies and early eighties the CFRJ underwent a great expansion, in addition to media exposure. This diffusion among a lay public formed by people not necessarily interested in becoming analysts (but who eventually did) was encouraged by a division that appeared around 1981, when the School started to offer two types of courses: graduate theoretical courses and clinical course aimed at training practicing analysts. The existence of graduate courses attracted a floating public gathered around the School and around Lacanian psychoanalysis. In 1982 and 1983 were held the famous "soirées,"[43] open encounters for the great public, to which prominent public personalities were invited.[44] We notice, from the choice of guests, an appreciation for irreverence, a search for estrangement from the culture considered as legitimate or academic, privileging, on the one hand, popular culture and, on the other hand, persons with a controversial profile or in-between popular and erudite cultural spaces. This type of choice was in accordance with M.D. Magno's project of producing a Brazilian psychoanalysis.

In this period the Freudian Schools of Vitória and of Brasília were also founded, united in 1983 under the aegis of the Freudian Cause of Brazil, which in the following year added the Freudian School of Goiânia.

In 1983 a new CFRJ statute designated M.D. Magno master of the school, meaning that reading Freud and Lacan should be accomplished through Magno's texts. An extremely controversial figure in the psychoanalytic milieu, Magno had in fact a very personal interpretation of the Lacanian work, proposing, as we saw, the production of a genuinely Brazilian theory. In time, and as it always happens, the excessive centralization around Magno, as well as his whims and idiosyncrasies, ended up causing significant dissidences. The ruptures preceded the extinction of the Cause in 1988.[45]

The second Lacanian institution founded in Rio de Janeiro had again the Argentine stamp. This was Freudian Letter, founded in 1981 by Eduardo Vidal, an Argentine psychiatrist invited to teach courses at SEPLA,[46] who ended by gathering followers in the institution for Lacanian training.[47] The foundation of the Freudian Letter was the beginning of a broad expansion of the Lacanian movement in Rio de Janeiro.[48]

Surprisingly, the proliferation of Lacanian societies took place in Brazil as a whole (although Rio de Janeiro remains as a record breaker), something that did not happen during the monopoly period of the "official" societies, not even at the peak of psychoanalytic diffusion. This proliferation outside the great centers, although dependent on them, can be explained by the saturation of the psychoanalytic market in these centers, which probably caused the new societies and psychoanalysts to search for followers outside the Rio-São Paulo axis, intensifying exchanges with groups from other states and enlarging the market, no so much of psychoanalysis itself, but of psychoanalytic training. Between 1982 and 1990, 15 Lacanian training societies were founded in the following cities: Brasília, Vitória, Salvador, Recife, Belo Horizonte (two), Curitiba (two), Teresina, Porto Alegre (two), Florianópolis, Fortaleza and São Luís (two). This information becomes even more surprising when we know that up to 1987 there were "official" psychoanalytic associations only in Rio de Janeiro (two), São Paulo and Porto Alegre, in addition to psychoanalytic centers in Recife and Brasília.

Psychoanalysis versus psychiatry: the present times

Nowadays we see a new form of diffusion of Lacanian psychoanalysis in the public institutions that provide care for low-income population, especially for those with severe psychiatric disturbances. Here I speak specifically about the Brazilian experience, because I could not obtain data on this subject in Argentina.

In Brazil, during the eighties, the Lacanian psychoanalysis boom shared space with three significant "psy" phenomena. One of them was the proliferation of "alternative" therapies, particularly Reichian therapies, or therapies derived of Wilhelm Reich's theory, that privileged working with the body to the detriment of spoken word.[49] Such therapies were complemented by divinatory practices (such as astrology or I Ching) and by parallel medicines (such as acupuncture or homeopathy), composing what I called at the time the "alternative circuit." An offshoot from the countercultural universe of the seventies, that group of practices competed (and, I imagine, compete still today) with psychoanalysis which, in its Lacanian version, opposed such practices vigorously, due to its theoretical (and scientific) strictness, and its fidelity to the spoken and written word. The study made by Lúcia Valladares de Oliveira shows a decrease in the São Paulo psychoanalysts' income during the eighties and nineties, at the same time that the profession became more "female," with a large inflow of women candidates that were also psychologists.[50] Even if this can be explained by the increasing number of psychologists graduated by the countless existent universities, another reason may have also contributed.

In the eighties, at the same time that psychoanalysis, whether Lacanian or not, began to feel the effects from competition with new alternative therapies, the psychiatric field underwent important changes. On the one hand, academic psychiatry was increasingly influenced by the biological (and pharmacological) trend coming from the United States. The great teachers gradually stopped teaching psychoanalysis, giving place to researchers linked to the pharmaceutical industry. Maybe this explains (at least in part) the "psychologizing" of psychoanalysis, which slowly ceases to interest young psychiatrists.

On the other hand, we see simultaneously the slow but firm ascension of the so-called "Brazilian Psychiatric Reform" movement in the realm of public policies. The "Brazilian Psychiatric Reform" is an eminently political movement, a heir of both anti-psychiatry (most of all in its Italian version) and of the leftist physicians' political fight during the military dictatorship. The RPB,[51] as it became known, ends by imposing an agenda of significant changes in mental health national policies and also at the local level.[52] During the nineties, the National Congress discussed the Paulo Delgado bill,[53] which, in its original version, prohibited the construction of new insane asylums in the country, foreseeing the gradual substitution of the existent facilities for substitute mental health services. In fact, an innovative policy of mental health financing led to the birth of Psychosocial Centers and Nuclei (CAPS and NAPS) in several states of the country, where psychiatric patients began to receive a type of care that excluded internment as an option. Furthermore, the presence of psychologists in health centers for low-income populations became increasingly expressive.

The existence of a psychoanalytic trend connected to Lacan's theory within the Brazilian Psychiatric Reform movement calls our attention. The work of Erotildes Leal, *O Agente de cuidado na Reforma Psiquiátrica Brasileira* (The caregiver in the Brazilian Psychiatric Reform) refers, for instance, to the way in which the movement presents a significant articulation with Lacanian psychoanalysis in Minas Gerais, a state which ranks just below Rio de Janeiro and São Paulo in the number of alternative mental care services.[54] A citation that appears in the text sheds some light in this matter:

For us, the issue was never whether we should or should not, but how to defend the theories of the Psychiatric Reform. For us, to say that the lunatic is a citizen is a corollary of The lunatic is a subject. In Minas Gerais, therefore, there is an alliance between Lacanian clinical practice and the Psychiatric Reform.[55]

As for psychoanalytic care in public clinics and health centers, it is also possible to see the influence of Lacanian psychoanalysis. In this case the Lacanian rejection of the strict rules of "official" associations, that prevented analysis outside the orthodox setting (couch, three to four times a week, sessions previously determined through a contract, etc.), favors exactly the conduction of psychoanalysis in any context, regardless of established settings. Similar to "the analyst authorizes himself," the other's "desire" is enough for an analysis to happen. The sessions may be scheduled each time, their duration may be variable, and the issue of a difficult communication with patients from popular strata tends to dissolve before the Lacanian type of "listening." It is enough to listen to what the other has to say, having the ability to deal with the unexpected and to value the absolutely unique path of each treatment. The patient from popular strata, in this sense, may even be a more interesting patient, since it is not polluted by the "psychologizing" so common in the middle class, being less influenced by preconceived ideas concerning what is psychoanalysis or what to expect from it, therefore able to speak freely – that is, freely associate – without trying to say what the analyst wants to hear. Therefore, the ideal of a newly reinvented psychoanalysis, as if it were experienced for the first time, is found more easily here.[56]

Lacanian psychoanalysis seems, therefore, to overflow the limits of private practice, reaching unorthodox patients and settings as seen from the standpoint of "official" psychoanalysis. This phenomenon certainly has multiple causes and motivations, of which I will try to point out a few. Initially, as Lúcia Valladares de Oliveira demonstrates, the psychoanalyst's clinic is no longer as profitable as it once was. This fact can be attributed to the competition of alternative therapies, but also and above all to the "biological" turning of psychiatry. The biological psychiatrists' current tendency is to leave aside their traditional role of caring for interned patients and occupy themselves in the profitable realm of the old neuroses – now transformed in "disorders", pharmacologically treatable in outpatient clinics and expensive offices. Turning to public mental health care – both in the case of the severely mentally disturbed and of the poor neurotic assisted in public health facilities – represents, therefore, the occupation of a space left nearly vacant by a psychiatry less and less interested in institutional policies and increasingly oriented to the politics of "scientific" production financed by pharmaceutical laboratories. Besides being a type of compensation (although not financial) for the ground lost in private practice, it is a positioning against the "pharmacological" psychiatry, through an alliance with the most progressive segments of the Psychiatric Reform, which tend to dominate the space of public policies. It represents the occupation of a significant space in the country's "psy" scenario, as well as the recovery of the most radical and political face of the movement.

Brazil and Argentina within the transnational Lacanian movement: shifting from the center to the periphery

In 1995 Roberto Harari and Isidoro Vegh, Argentine psychoanalysts, planned in Buenos Aires to gather a group of international Lacanian associations. This initiative gave rise, in 1998, to the foundation of the Convergence – Lacanian Movement for Freudian Psychoanalysis, with articles of incorporation drawn in Paris by 13 French associations, two Italian, four Spanish, one German, one American, one Uruguayan, 14 Argentine and eight Brazilian.[57] Taking into account the origin of the founding institutions, Spanish seems to have surpassed French as the language of psychoanalysis. If we add Portuguese, we see that in fact today there is a significant geographical shift, similar to what happened when English substituted German as the preferential language for the transmission of psychoanalysis.

It is important to note that the Convergence was founded by Harari and Vegh as a counterpoint to the World Association of Psychoanalysis (AMP), a sort of Lacanian IPA founded under the leadership of Jacques Alain Miller.[58] Is the Convergence then merely a Latin-American, or local, opposition to AMP, without importance in the transnational Lacanian scenario? This does not seem to be the case. The AMP, as Renata Pereira tells us, emerged from a Pact signed in Paris in February 1992 by four schools, the School of the Freudian Cause (France, 1991), the Caracas School of the Freudian Field, (Venezuela, 1985), the European School of Psychoanalysis of the Freudian Field (France, 1990) and the Lacanian School of the Freudian Field (Euro-Argentine, 1992).[59] In other words, also in the "enemy" field the alliance between France and Latin America was in force. When we observe the places where AMP's national meetings and Congresses were held, the alliance seems somewhat more unbalanced, decidedly tilting to the Argentine side:

- IX International Meeting, held in 1996 in Buenos Aires.
- AMP's 1st Congress and 10th International Meeting, 1998, in Barcelona.
- AMP's 2nd Congress and 11th International Meeting, 2000, in Buenos Aires.
- AMP's 3rd Congress and 12th International Meeting, 2002, in Brussels-Paris.
- 1st American Meeting of Freudian Field Schools, 2003, Buenos Aires.
- PIPOL Meeting (International Research Program on Lacanian Psychoanalysis) – European, 2003, Paris.
- AMP's 4th Congress, 2004, Comandatuba (Bahia).[60]

The current Lacanian movement, therefore, seems to be French-Argentine, with strong roots in Brazil,[61] tending to spread to other Latin America and Caribbean countries.[62]

More than a simple propagation of the Lacanian psychoanalysis segment beyond French borders, I believe that we are witnessing a new decentralization of psychoanalysis, similar to the one which took place in the forties with the 2nd World War and, later, in the seventies, with the ascension of Lacanism (and the simultaneous decadence of American psychoanalysis). This third shift, as each of the former ones, has its own characteristics. The most outstanding is, undoubtedly, the fact that this is a shift from the center toward the periphery. That this displacement was welcomed and encouraged by Lacanism, I think there is no doubt. The explanation I propose for such a fact is a double one. On one side, as we saw, in countries such as Brazil or Argentina, psychoanalysis was highly valued in a moment of great political repression and social oppression, as an instrument of liberation.[63] The Lacanian segment, in its strong opposition to constituted authority (IPA), gave a voice to the anti-establishment feeling that already fueled an intense diffusion of psychoanalysis in the "psy" field of both countries. On the other hand, Lacanism's inner dynamics, as we also saw, in valuing the charismatic form of transmission control favors dispersion and dissidence, and therefore transnationalism. Those are some of the facets of this unusual decentralization that, in its complexity, will certainly produce other interpretations and forms of understanding.

The fact is that, changing with the centrifugal force of its own propagation movement, psychoanalysis remains, in the early twenty-first century, as a sociological mystery to be revealed. Its ability to act as a world view and to make sense of people's lives seems to challenge the dark forecasts that for a long time announced its decline. Above all, it challenges cursory analyses that, focusing the hegemonic centers of scientific diffusion, do not see the great ebullience taking place in the periphery.

Notes

1. Also known as biological psychiatry, which offers a physical interpretation for mental disorders, completely based on neurochemical mechanisms.
2. Acronym of the International Psychoanalytical Association.
3. See Max Weber, *Economy and Society*, ed. Guenther Roth and Claus Wittich (Berkeley: University of California Press, 1978), pp. 956–1158.
4. Largely based on Mariano Ben Plotkin's book *Freud in the Pampas* (Stanford: Stanford University Press, 2001).
5. It is true that I have a few excuses on my behalf, namely Brazil's extension (both territorial and populational) as compared to Argentina, the existence of at least two significant diffusion centers (Rio de Janeiro and São Paulo) and, particularly, the well-known proliferation of groups and institutions in Rio de Janeiro. All these factors tend to complicate the analysis of the Brazilian case.
6. See Elizabeth Roudinesco, *História da Psicanálise na França – A Batalha dos Cem Anos*, Volume I: 1885–1939, (Rio de Janeiro: Jorge Zahar Editor, 1989), pp. 131–138.
7. Despite the opposition of the European psychoanalytical movement and of Freud himself, the North-American psychoanalysis societies, since their inception, would accept only M.Ds as candidates. In England, this rapprochement to medicine took place through a sort of "medical" reading of Freudian texts.
8. In addition to medicalization and to pragmatism, there was a certain adaptation of the doctrine promoted by immigrant psychoanalysts who, after the traumatic experience of Nazi persecution and an extremely bloody war, resorted to a more optimistic and healing version of the Freudian theory.
9. I would like to point out that IPA never issued resolutions prohibiting non-doctors' access to psychoanalytical training.
10. The word "transmission" distances itself from IPA's more pedagogic terminology (such as "didactic analyst" or even "supervision") which puts the candidate in the position of a student.
11. As well as Argentine and Brazilian associations, as we shall see.
12. About the social diffusion of psychoanalysis in Argentina during the twenties and thirties, see Mariano Ben Plotkin, Freud, politics and the Porteños: the reception of psychoanalysis in Buenos Aires, 1910–1943, *Hispanic American Historical Review*, 77 (1997): 45–74. About the same process in Brazil, see Jane Russo, A difusão da psicanálise no Brasil na primeira metade do século XX – da vanguarda modernista à rádio-novela, *Estudos e Pesquisas em Psicologia*, 2 (2005): 53–64 and Roberto Sagawa, "A Psicanálise pioneira e os pioneiros da psicanálise em São Paulo" in *Cultura da Psicanálise*, ed. Sérvulo Figueira (São Paulo: Brasiliense, 1985), pp.15–34.
13. From the end of the thirties Arnaldo Raskovsky and Enrique Pichon Rivière, both medical doctors, led a study group on Freud's works. In 1938 Angel

Garma arrived in Buenos Aires, a Spanish psychiatrist with a psychoanalytical degree from the Berlin Institute. Then Celes Cárcamo arrived, an
Argentine psychiatrist who had graduated in Paris. The two, together with
the group led by Rascovsky and Pichón Rivière, founded the Argentine Psychoanalytical Association. Therefore, this was not about "IPA's people." In
the case of Garma, the fact of being Spanish made him nearly a local. See
Plotkin, *Freud in the Pampas*, pp. 44–47.
14. When the Buenos Aires Psychoanalytical Association was founded (APdeBA).
15. See Plotkin, *Freud in the Pampas*, pp. 204–214.
16. See Sergio Visacovsky's contribution to this volume.
17. In São Paulo, the then young psychiatrist Durval Marcondes founded in
    1927 the Brazilian Psychoanalytical Association (Sociedade Brasileira de
    Psicanálise), which acquired a Rio de Janeiro chapter in the following year
    and obtained temporary acknowledgement by IPA in 1929. The association
    attained great success among the São Paulo elite, its meetings being attended
    by artists and intellectuals, as a social event. It was finally terminated after
    a brief period, since, with the exception of Durval Marcondes, there was no
    real interest in psychoanalysis training among the attending public.
18. See Carmem Lucia Valladares de Oliveira, *História da Psicanálise – São Paulo
    (1920–1969)* (São Paulo: Escuta, 2006).
19. The instability that characterized the beginning of the Rio de Janeiro "offi-
    cial" movement remained until recently. In 1999, the Rio 3 Psychoanalysis
    Association, a dissidence of SBPRJ, was authorized by IPA to function as a
    study group and is now an officially recognized association. In 2002, the
    Association of Psychoanalysis of the State of Rio de Janeiro, dissidence of
    SPRJ, was authorized to work as a temporary association.
20. In Argentina, the period that ranges from the military coup of 1966 until
    1983 is one of intense repression and State terrorism. As of 1970 the left-
    ist parties dramatically radicalized their politics, financing their actions
    with kidnapping and armed robberies. Meanwhile, the Argentine Anti-
    Communist Alliance (Triple A) had Isabelita Peron's support in their repres-
    sion actions. Political violence, therefore, started before the military coup
    that deposed Isabelita in 1976. Between May 1973 and March 1976 there
    were 1207 people dead and 847 wounded in the "people's side" and 336 dead
    and 604 wounded in the "regime's side." See Angelo Priori, Golpe militar na
    Argentina: apontamentos históricos, *Revista Espaço Acadêmico*, 59 (2006). In
    Brazil the 1964 military coup only acquired the outlines of a violent and
    repressive dictatorship in December 1968. During the first half of the sev-
    enties there was a period of intense repression and political shutdown, with
    the formation of urban and rural guerrilla groups that, however, did not get
    the expected support from the population. After a "slow, gradual and pro-
    gressive" period of opening that resulted in the 1979 Amnesty Act, a civil
    president was elected by Congress in 1985.
21. To the contrary, the case of Amilcar Lobo, that surfaced during the eighties,
    showed how a SPRJ candidate that was a military medical doctor working
    with torture groups (to determine to what extent the tortured could tolerate
    the "treatment" inflicted) was covered up by his didactic-analyst and his
    supervisor and, according to accusations that were "hushed" in the seventies,

by the high administration of the association. See Cecília Coimbra, *Guardiães da Ordem* (Rio de Janeiro: Oficina do Autor, 1995), pp. 94–106.

22. The Institutional Act no. 5 (AI-5) was issued by the military government in December 1968. Considered as a coup inside the coup, AI-5 extinguished what was left of Brazilian democracy, giving rulers the power to persecute, arrest and suspend all the political rights of enemies of the regime, in addition to determining the immediate recess of the National Congress. The Act was revoked only 10 years later. The term AI-5 Generation, therefore, refers to those who, being young at the time, did not take part on the intense political efervescence before 1968.

23. The best-known analysis from this standpoint is the one by Luciano Martins *A Geração AI-5* (The AI-5 Generation) published in *Ensaios de Opiniao*, 11 (1979), pp.72–102. However, the idea is taken up again in several texts that analyze the period.

24. On this period, see Heloísa Buarque de Hollanda, *Impressões de viagem – CPC, vanguarda e desbunde, 1960/70* (Rio de Janeiro: Aeroplano, [1979] 2004); Carlos Alberto Pereira, *Retrato de Época – a poesia marginal nos anos 70* (Rio de Janeiro: FUNARTE, 1981) and Gilberto Velho, *Nobres e Anjos, um estudo de tóxicos e hierarquia* (Rio de Janeiro: Jorge Zahar Editor, [1975] 2005).

25. I develop this argument in my book *O Corpo contra a palavra – as Terapias Corporais no campo psicológico dos anos 80* (Rio de Janeiro: Editora UFRJ, 1993).

26. We have already seen that this was not the case of the São Paulo Association. However, the work of Lúcia Valladares de Oliveira shows that non-medical doctors were minority until the eighties, maybe indicating a type of selection that privileged medical candidates.

27. On the wake of the opening of new psychology courses, during the sixties, the number of professionals graduated in Brazilian soil grew exponentially. In 1974, when the Federal Council of Psychology started operations, there were 895 enrolled professionals. In 1975 the number rose to 4.950 and the following year to 6.890. See Deise Mancebo, *Da psicologia aplicada à institucionalização universitária: a regulamentação da psicologia enquanto profissão*, *Cadernos IPUB*, 8 (1997), pp.161–177.

28. They were Mauricio Knobel, Eduardo Kalina, Arminda Aberastury, Léon Grinberg and Arnaldo Rascovsky.

29. Lobo invited Arminda Aberastury and Eduardo Kalina for seminars on childhood and adolescence, proposing to extend them to the outside public. His proposal was vetoed by the direction of the Association. Lobo, who supervised a large group of psychologists working in child care, formed a parallel institution (Institute of Psychological Orientation-IOP) that started to provide courses to psychologists. In addition to Aberastury and Kalina, several other Argentine psychoanalysts, most of them linked to ASAPPIA, gave courses at IOP. Starting with IOP members, the Society of Clinical Psychology was founded in 1971 (currently Psychoanalysis Society of Rio de Janeiro city). See Ana Cristina de Figueiredo, *Estratégias de difusão do movimento psicanalítico no Rio de Janeiro 1970–1983* (Master diss., PUC-RJ, 1984).

30. According to the testimony of APPIA's last president to Ana Cristina Figueiredo, the institution was a sort of Argentine embassy. In fact, APPIA

"distributed" the Argentines arriving in Brazil, whether temporarily or definitively, among several states, depending on the degree of demand for courses and graduation. This was the case of Vitória, Espírito Santo's capital, where the psychoanalytical movement started with the arrival of the Argentine psychoanalysts sent by APPIA. See Cíntia Avila de Carvalho, *Os psiconautas do Atlântico sul, uma etnografia da psicanálise* (Ph.d diss., Universidade Estadual de Campinas, 1995).

31. In Ana Cristina Figueiredo, *Estratégias*, pp. 45–46.
32. Manoel Tosta Berlinck, Difusão e construção, in *Freud 50 anos depois*, ed. Joel Birman (Rio de Janeiro: Relume Dumará, 1989), p. 71.
33. Ibid., p. 72. There are at least two other important figures of the "Argentine invasion" worth mentioning. One is Gregorio Baremblitt, who founded in Rio de Janeiro the IBRAPSI, Brazilian Institute of Psychoanalysis, Groups and Institutions, with the purpose of furthering the proposal of Mental Health Care Workers, in their Marxist perspective. The other is Emilio Rodrigué, one of the most eminent among the second generation of APA psychoanalysts, who left the Association with the Document and Platform groups in 1971, changing dramatically his career. Rodrigué, once president of the APA began to experiment with different types of alternative therapies. In his many trips to Brazil, he helped to introduce the new body therapies among young professionals. Settling finally in Salvador (Bahia) he is now considered to be a source of inspiration for the Bahia psychoanalysts. Owing to Rodrigué's unorthodox career, the psychoanalytical scene in Bahia is quite original, as the baiano "psy" professionals were first trained in bioenergetics and reichian therapy, before embracing lacanian psychoanalysis.
34. The "official" associations, both in Rio and São Paulo, followed the English school, adhering essentially to Melanie Klein's theory, with a certain degree of Bion's influence in São Paulo.
35. Published as *Agenda of Psychoanalysis* in 1989 and 1990 and as the *Brazilian Yearbook of Psychoanalysis* (*Anuário Brasileiro de Psicanálise*) in 1991 and 1992, it had the purpose of mapping the psychoanalytic field in Brazil and to disseminate the several activities of the different associations.
36. Daniela Ropa and Denise Maurano ed., *Agenda de Psicanálise* (Rio de Janeiro: Relume Dumará, 1990), p. 56.
37. Ibid., p. 150.
38. Ibid., pp. 19–20.
39. About the contradiction between a highly enigmatic language and its diffusion, Sérvulo Figueira remarks: "Lacan's style – that is, his personal style – entangles the reader in a plot of thoughts from which he will hardly escape, since its obscurity throws the reader in the position of a disciple who does not know and who desperately seeks the meaning of the words of an all-knowing Master. The reader becomes the analyzed, since what is left to him, besides giving up, is to transfer. And the Master's speech, supposedly impenetrable to prevent easy dilution, spreads as no psychoanalysis ever could, exactly because not being understood, it settles comfortably in the speech as a slogan and cliché, to be used according to the situation. (Sérvulo Figueira, Quem tem medo de Jacques Lacan, in *Os Bastidores da Psicanálise. Sobre política, história e estrutura do campo psicanalítico* (Rio de Janeiro: Imago, 1991), p. 49).

40. In 1988, Forbes founded, together with 17 Library members, the São Paulo Psychoanalytical Association, in an attempt to gather psychoanalysts from varied groups around a discussion on Lacanian psychoanalysis.
41. Manoel Tosta Berlinck, *Difusão*, p. 74. In fact, in the four issues of the Psychoanalysis Calendar and Yearbook, 14 Rio de Janeiro institutions were listed as having pronounced themselves Lacanian, and only five from the São Paulo institutions listed could be so considered. Although the Agenda listings cannot be considered exhausting (since they depended on associations adhering to the project), the numbers indicate higher vitality in the Rio de Janeiro Lacanian movement.
42. According to Ana Cristina Figueiredo, the School was founded in 1976, which makes sense, since M.D. Magno met Betty Milan in Paris only in 1975.
43. The portuguese word is sarau, which means a private party given at night, usually with a literary of musical purpose.
44. Singer Caetano Veloso was the first one. After him there were Carnival scenographer Joaosinho Trinta, poet Ferreira Gullar, the staff from *Casseta Popular* and *Planeta Diário*(humor publications), the legendary host of TV shows Abelardo Chacrinha Barbosa, theater director José Celso Martinez Correa, known by his *avant-garde* shows, among others.
45. Charismatic control, in the absence (or scarce presence) of "rational" bureaucratic regulations limiting fidelity based merely in the extraordinary qualities of a person, always has a price to pay: successive dissidences and ruptures often resulting in new institutions, in a continuous updating of charisma under a new master.
46. Latin-American Psychoanalytic Studies Association, founded in 1978.
47. Roberto Harari is another Argentine linked to the diffusion of Lacanian psychoanalysis in Brazil, having taken part in the foundation of two associations in the south of the country – Mayeutica in Porto Alegre and Miaêutica in Florianópolis.
48. Between 1983 and 1998 13 Lacanian (or soon-to-be Lacanian) associations were founded in the city of Rio de Janeiro.
49. See Russo, *O corpo*.
50. See Carmem Lucia Valladares de Oliveira, História, pp. 267–277.
51. Which stands for Reforma Psiquiátrica Brasileira.
52. As in the paradigmatic experiences carried out in the cities of Santos (State of São Paulo) and Angra dos Reis (State of Rio de Janeiro), both under the administration, at the time, of the Workers Party.
53. Federal bill prepared by the leadership of the psychiatric reform movement and submitted to the House of Representatives in 1989 by Congressman Paulo Delgado. After 12 years in the Congress, the project was voted and approved in 2001. This federal law – 10.216 – preserves the spirit of the original text, determining preferential treatment in community mental health services, replacing therapeutic isolation by social reinsertion. Furthermore, during the nineties, several Brazilian states – such as Ceará, Pernambuco, Rio Grande do Norte, Federal District, Minas Gerais, Paraná and Rio Grande do Sul – approved laws inspired in the Paulo Delgado project on behalf of the progressive replacement of hospital internment by outpatient services.
54. It should be pointed out that, in addition to a more radical and political reading arising from the Italian reform, the Brazilian Psychiatric Reformation

also received strong influence from French institutional psychiatry, intensely based in Lacan's psychoanalysis.

55. Excerpt from a book by F.P. Barreto entitled *Reforma Psiquiátrica e Movimento Lacaniano* (Psychiatric Reform and Lacanian Movement) mentioned in Erotildes Leal's "O Agente de cuidado na Reforma Psiquiátrica Brasileira, (Phd. diss, Universidade Federal do Rio de Janeiro, 1999), p. 124.

56. This rapprochement between Lacanian psychoanalysis and public mental health services may be seen in the works by Ana Cristina Figueiredo and Sonia Alberti ed., *Psicanálise e Saúde Mental: uma aposta* (Rio de Janeiro: Companhia de Freud, 2006); Ana Cristina Figueiredo, *Vastas Confusões e Atendimentos Imperfeitos: a clínica psicanalítica no ambulatório público* (Rio de Janeiro: Relume Dumará, 1997). Maria Tavares Cavalcanti and Ana Teresa Venâncio ed., *Saúde Mental: campo, saberes, discursos* (Rio de Janeiro: Edições IPUB-CUCA, 2001); Fernando Tenório, *A psicanálise e a clínica da reforma psiquiátrica* (Rio de Janeiro: Rios Ambiciosos, 2001), among others.

57. See Renata Susan Pereira, Formação e Instituição : um percurso pela história das instituições psicanalíticas de Florianópolis (Master's diss.: Universidade Federal de Santa Catarina, 2005), p. 119.

58. Son in law of Jacques Lacan, becomes a controversial figure in the psychoanalytical scenario for trying to retain a strict control of Lacan's legacy (by controlling both the institution founded by Lacan and the publication of his last seminars).

59. The Brazilian School of Psychoanalysis, affiliated to AMP, which is actually the transformation of certain preexistent associations into one unified school, was formally founded in 1995 in a meeting held in Rio de Janeiro. According to the model proposed by AMP, the School maintained representatives in several Brazilian states, among which (in addition to São Paulo, which ended up by becoming its headquarters), Bahia Paraná, Santa Catarina and Minas Gerais. (See the above-mentioned work by Renata Susan Pereira). A Pereira informer refers to groups in Bolivia, Venezuela, Cuba and Colombia, which were not yet schools.

60. Renata Susan Pereira, *op.cit.*, p. 150.

61. I had no information on Mexico, which is also apparently located within the Lacan school's area of influence.

62. Spain is an interesting case, because although I lack information on Lacanian Spanish associations and their foundation, it is possible to guess that the Spanish movement is subsidiary to the Argentine movement, particularly when we think that Oscar Masotta, founder of the first Lacanian institution in Argentina, was responsible for the introduction of the Lacanian theory among Spaniards.

63. Whether of political liberation, as in the Argentine case, or personal liberation, as in the case of Brazil.

9
Origin Stories, Invention of Genealogies and the Early Diffusion of Lacanian Psychoanalysis in Argentina and Spain (1960–1980)

Sergio Eduardo Visacovsky

As with most expert knowledge, psychoanalysis is thought of by followers as a theory and clinical practice with universal reach, immune to the effects that the specific characteristics of national, regional or urban environments can cause. The differences, it is maintained, arise from the different interpretations of Freudian theory produced by the psychoanalytical community, or from the personal characteristics of analysts and patients that can generate different styles of treatment or transmission. Against these wishes, studies on the transnational diffusion of consumer goods, lifestyles, cultural products, knowledge, images and beliefs have shown the crucial importance that local conditions of reception. In other words, the diffused and received objects do not remain unaltered; instead they are necessarily accepted and provided with a sense according to the local predominant types of cultural interpretation.[1] The types of knowledge defined as "expert" – like psychoanalysis – do not escape the same rule even though their legitimate cognitive aims are universal. They are, above all, social practices rooted in cultural traditions and networks of meanings,[2] placed, interpreted and appropriated in singular contexts.

Much of the research focusing on analyzing the processes of the diffusion of psychoanalysis around the globe have highlighted, precisely, the importance that the social and cultural conditions of reception posses. For example, the favorable acceptance that psychoanalysis has had in the United States and its transformation into "Ego Psychology" has been explained through the way in which it was reinterpreted through philosophical traditions like pragmatism, deeply rooted in the North

American society, which encourages values like optimism and individualism, transmitted by the main receiving and diffusing agents, the doctors. Meanwhile, in France, psychoanalysis was the initial object of interest of artists and writers, discovering resistance in the medical community, in institutions such as the Catholic Church and in political-philosophical currents like Marxism.[3] In sum, psychoanalytical ideas are not received, appropriated and applied in a historical, social and cultural void, but established according to preexisting particular conditions. Undeniably, the contextual peculiarities can explain theoretical and therapeutic acceptances and resistances; these singularities can refer to the medical-psychiatric, psychological, philosophic or literary fields, the institutions and political ideas, intellectual traditions and religious beliefs, and finally to the dominant concepts around gender, sexuality, the person, the body, the family, the mind, pathology or madness.

The diffusion and reception of a psychoanalytical theory and practice produced in a specific national context where a previous insertion of psychoanalysis exists represents a very particular case of transnational diffusion. For chronological reasons, this has been the case of the theories of the French psychoanalyst Jacques-Marie Émile Lacan (1901–1981), which began to circulate outside France in the early 1960s. The general aim of Lacanian perspective was "the return to Sigmund Freud", that is to say, to question the versions adopted by the International Psychoanalytical Association (IPA). Lacanism viewed the IPA's versions as distorted or partial, and sought to re-orient psychoanalysis along the path taken by its creator, Freud. Lacanian theory was transmitted in different national contexts with diverse degrees of institutional development of psychoanalysis and the acceptance of psychoanalytical ideas on the part of the society. In this sense, Argentina and Spain have represented peculiar cases of diffusion of Lacanian psychoanalysis.

Lacan's ideas were introduced in Argentina at the beginning of the 1960s, through the influence of French thought among intellectuals. Lacan´s thought met a vigorous psychoanalytical field already constituted since the early 1940s around the Asociación Psicoanalítica Argentina (APA), institution officially affiliated to the IPA, and an outstanding level of insertion of psychoanalysis not only in the medical world but also in various disciplines and even in daily life. In contrast, Lacanian concepts were established in Spain in the mid-1970s, within the frame of a weaker psychoanalytical field. In Spain, psychoanalytical institutionalization began shortly after 1959, in the context of a long history of rejection and peculiar interpretations of Freudian theory by psychiatrists and philosophers. What is remarkable about the case is that the diffusion of Lacan's ideas in Spain came from Argentina, and not directly from France; that is to say the diffuser agents came from a psychoanalytical context then considered peripheral in comparison to the European and North American centers; and that the principal transmitter agent was the same person attributed with the introduction and spreading of Lacan's ideas in Argentina: the writer and essayist Oscar Masotta (1930–1979). Born in Buenos Aires, he did not come from the medical field nor did he complete his academic studies at the Universidad de Buenos Aires (UBA). He trained himself auto-didactically in theoretical currents like phenomenology and structuralism, and in disciplines like literary criticism (he published the book *Sexo y traición en Roberto Arlt* in 1965), the semiotic (in particular, applied to the problematics of architecture and urbanism), modern art and aesthetic vanguards and comic strips. He had contributed to *Contorno*, one of the most important journals of the so called "new intellectual left". Lacan was mentioned by Masotta already in 1959,[4] in 1964 he gave a lecture on "Jacques Lacan and the contemporaneous French psychoanalysis"; and in 1965, Masotta published his first article on Lacan in *Pasado y Presente*, the journal of dissidents from the Communist Party.[5] But it was not until 1969 that Masotta took on the task of organizing study groups focusing on Lacan's ideas.[6] In 1974 he founded the Escuela Freudiana de Buenos Aires, the first Lacanian institute in a Spanish-speaking country. At the end of this year Masotta left Argentina for London and then went to Barcelona, where he founded the Biblioteca Freudiana in 1977, the first Lacanian institution in Spain. He was an active diffuser of Lacan's ideas in Spain until 1979, when he died.

According to what has been pointed out, the contextual differences between Argentina and Spain produced dissimilar conditions of reception and diffusion of Lacan's thought. How the newly spread currents, the diffuser agents and the established psychoanalytical institutions gained legitimacy has not been widely studied. Along the course of psychoanalytical history, the theoretical and institutional disputes, crowned on more than one occasion with ex-communions and splits, invoked presumptions of pureness in the face of rivaling positions considered deviations of Freudian or spurious forms. The representations of psychoanalytical pureness make up, simultaneously, a way of inscribing theories, people and institutions in a universal and transnational history of psychoanalysis, and a way of affirming the genealogical continuity with Freud as the father of psychoanalysis.[7]

The introduction of Lacanism in countries like Argentina and Spain supposed the constitution of new groups of psychoanalytical families, based on the pretence of descending from Lacan (and through him, from Freud). Moreover, at the same time, the genealogies should not only be projected in respect to the shared psychoanalytical ancestors on an international scale, but should also establish local affiliations. That is to say, those who were the authentic descendants of Freud and Lacan in each national context, and for that motive, could be considered fathers of a Lacanian genealogy in a local context. In the cases of Argentina and Spain, Masotta would assume the role of father of a new psychoanalytical genealogy; but the ways in which the relationships between Masotta and local psychoanalytical ancestors were established, the way in which the principles that secured the admission of psychoanalysts and institutions as legitimate descendants of Masotta were instituted, were notably different.

This article has the general objective of presenting a case of transnational diffusion of psychoanalysis, from a country described as peripheral in the 1970s – Argentina – which nonetheless introduced a psychoanalytical current, Lacanism, into an European national context, which was marginal from the psychoanalytical development – Spain. The exact point of our analysis consists in showing the specific forms in which the origins of Lacanism were inscribed in the psychoanalytical histories in Argentina and Spain. In Argentina, the action of diffusion of Lacan was based largely on the struggle against the power established by the APA and psychoanalysis in hands of the doctors. For its legitimation, Lacanism instituted genealogical relationships with a part of the psychoanalytical local past. This implied recognizing that psychoanalysis in Argentina existed before the diffusion of Lacan's ideas, and that part of the preexisting psychoanalysis was acceptable, because the tradition inaugurated by Masotta would be seen, to a certain point, as a continuity with the past. In Spain, after Franco's dictatorship, a democratic space opened up favoring the diffusion of discourses that had been prohibited up until then. But in relation to the diffusion of Lacan's thought, difficulties emerged for the construction of a Lacanian psychoanalytical genealogy in Spain. Was it possible to look for local genealogies and ancestors, like in Argentina? Could ideas and practices defined as "psychoanalytical" in the past be considered as "psychoanalytical" in the present? If there was an affirmative answer, how could the existence of psychoanalysis under Franco's authoritarianism be explained? If there was a relationship between a preexisting psychoanalysis and the one brought by Masotta, how could the qualities of Lacanian psychoanalysis be defended as new and revolutionary? As we will see, the peculiar aspect in this process of re-elaboration of the past was that

Lacan's ideas were diffused from France to Argentina and from Argentina to Spain together with specific narrations about Masotta's actions of transmission. Together with these stories, assumptions of genealogical continuity were diffused, a fact which would differentiate the legitimate descendants of Masotta from those who were not.

In sum, this work intends to analyze how the narrations of the early diffusion of Lacanism in Argentina and Spain by the same agents implied a process of inscription and conciliation of the same – as an origin story – regarding particular established forms of narrating the national psychoanalytical past and how this task involved a global reinterpretation of each psychoanalytical past. As we will see, this had the main objective of inventing genealogies that could give legitimacy to the individuals, and new Lacanian institutions founded in the name of psychoanalysis.

The origin stories of Lacanism in Argentina

A very well known story about the origin of Lacanism in Argentina has been circulating among local psychoanalysts. The story tells how the famous Argentine psychoanalyst Enrique Pichon Rivière gave Oscar Masotta the mimeographed manuscripts of Lacan. Pichon Rivière was born in Geneva in 1907, and died in Buenos Aires in 1977. He arrived in Argentina when he was 3 years old. He worked as a journalist, and played an active part in the porteño intellectual Bohemia. He qualified as a physician in 1936 at the UBA, and later he was appointed psychiatrist at the Hospicio de las Mercedes, the public mental hospital for male patients in Buenos Aires. Pichon Rivière coordinated there significant changes in the psychiatric treatments during the 1940s (from the application of electroshock in Argentina to the introduction of group therapy techniques). A reader of Freud's works since his formative years, Pichon Rivière was one of the founders of the APA in 1942, together with Celes Cárcamo, Ángel Garma, and Arnaldo Rascovsky. He also actively participated in the reorganization of public mental health care in the 1950s, while he was unusually fond of poetry, literature, art, philosophy and soccer. Pichon Rivière founded the Escuela de Psiquiatría Social in 1953, where Masotta gave his celebrated lecture about Lacan in 1964. Since the 1960s Pichon Rivière was inclined toward a psychoanalytical perspective influenced by Marxism and other sociological and cultural anthropological currents. At that time in Argentina's psychoanalytical field, Pichon Rivière was characterized for being, perhaps, one of the few official psychoanalysts who before the 1960s had explicit social and political concerns. As we will see, Pichon Rivière criticized the psycho-analytical dogmatism and a-political position of the APA's members, and thus he supported the groups which split off from the APA in the early 1970s.

As I mentioned, the story tells of a meeting between Pichon Rivière and Massotta. The tale supposes that Pichon Rivière gave these works to Massotta based on the fact that he (Massotta) might have been deeply interested in them and counting on his intellectual capacity to under-stand them. This story has several versions,[8] and the oldest of the stories comes from Massotta's own personal account.[9] The versions have mini-mum differences: all assume a relationship between Pichon Rivière and Massotta; all the versions affirm that the works of Lacan were in Pichon Rivière's control; all suppose some kind of relationship between Pichon Rivière and Lacan (some emphasize that they had a close relationship of mutual respect, justifying the gift to Pichon Rivière); all of them agree in emphasizing Pichon Rivière's decision to give away such a precious treasure, and to give it over to Massotta; finally, some versions emphasize that Pichon Rivière gave the texts to Massotta because he was "generous", but also because of his incomprehension and disinterest for Lacan's thought.

Perhaps one of the most emphasized aspects of the story is the assumption that Pichon Rivière had a misunderstanding or disinterest of Lacan's manuscripts, since they explain (along with his generosity) the gift to Massotta. The story tries to give an answer to clarify why such a notable psychoanalyst and intellectual as Pichon Rivière had on his bookshelf manuscripts from Lacan. According to Massotta, the writings were a very valuable treasure, but nonetheless, Pichon Rivière did not seem to value them as much as Massotta did. In other words, how could these writings have fallen into the hands of someone who did not give them the value that they certainly had? And, how could someone recog-nized for his values as a psychoanalyst and intellectual not have given the value to the manuscripts he held in his bookshelf?

Pichon Rivière himself gave plausibility to the story, since he con-firmed his relationship with Lacan. Pichon Rivière defined this rapport as "friendship" (the manuscripts were actually dedicated to Pichon Rivière by Lacan), which stemmed from a French speaking psychoan-alytical conference in 1951 and grew on the basis of a common interest to renovate psychoanalysis through a reinterpretation of Freud. They both also shared a love for surrealism and French literature. Nonethe-less, Pichon Rivière assured that he had differences with the Lacanian focus which he characterized as "idealist" for failing to recognize the fundamental determination that social relations play in the constitution of the subject (according to the principles of Marxism).[10] We do not know if these opinions expressed his early thoughts about Lacan, or if they would be better situated in the 1970s. Whatever it was his authentic reason, Pichon Rivière, the possessor of Lacan's manuscripts in Argentina, had decided to give them over to Masotta, making him his legitimate owner.

Other versions suggest that some of Masotta's first disciples convinced him to start organizing study groups focusing on Lacan's works.[11] At the same time, these versions emphasize that Lacan's name already circulated in Buenos Aires during the early 1960s; especially, it was mentioned by the French philosopher Louis Althusser. Many philosophers, writers and psychoanalysts already knew about the existence of Lacan's writings, fact which shows that attributing the discovery of Lacanian theory solely to Masotta is far fetched. The stories, while putting emphasis on how Masotta could have received the Lacan´s works would provide us with the understanding that Masotta turned into the diffuser of Lacan in Argentina. However, can one personal meeting, alone, explain the entrance and diffusion of Lacan in Argentina? Do these stories give case examples which help us to understand why Lacanian ideas became an object of interest and obtained a predominant place in Argentine psychoanalysis in the 1980s? From our perspective, the objective is not to refute versions of the stories or compare them to what happened in reality; on the contrary, we think that it is essential to rebuild the intellectual and psychoanalytical fields shaped in the late 1950s and the late 1960s. The stories refer to this moment in time, which was constituted when the first diffusion and reception of Lacan in Argentina was produced.

The diffusion of Lacanism in the intellectual and psychoanalytical field in Argentina (1955–1970)

During the Peronist government (1946–1955), public universities such as the UBA were controlled by nationalists and conservative Catholic intellectual groups. As a consequence of this, liberal, socialist and communist intellectuals were marginalized or expelled from the universities. When Perón was overthrown, the situation was inverted: intellectuals excluded in the precedent period returned to the academic institutions, whereas a big part of the others either left the universities on their own free will, or they were forced to resign to their university positions. Consequently, the intellectual field was organized around the political identity of Anti-Peronism. This was the context in which Lacan's ideas were received and diffused in Argentina for the first time.

The intellectual world was organized around the ideal of a "committed" intellectual that by then seemed deep-rooted in the work of the French philosopher Jean-Paul Sartre, allowing the consolidation of an intellectual block formed from liberal and leftist currents. However, this alliance began to dissolve during the course of the 1960s, with the Cuban Revolution, the surfacing of Latin American and Argentine armed guerrilla foco groups, the increase of repression against the groups identified as revolutionaries, and also the reinterpretation of Peronism that began to be appropriated by the Left. The organizing role of Marxism allowed the transition from an intellectual adhering to a "commitment" to that of the "revolutionary intellectual". This new intellectual model acquired hegemonic dimensions after the June 1966 military coup in Argentina and the successive historic events which allowed a candidate representing Peronism to win the presidential elections in 1973.

In this context, the existentialist humanism of Sartre was yielding a space for the reception of the novel intellectual trends, among which stood French structuralism.[12] As the old intellectual agreement began to break from liberalism, the intellectual field (auto defined as "progressive") restructured itself around Marxism. This turn to Marxism was not the only system of ideas that allowed this newly formed "leftist" intellectual field to grow, but also permitted some authors and intellectual currents that up until then were considered marginal to obtain legitimacy. Psychoanalysis was among these intellectual currents which gained legitimacy. Such was the case of Althusser, who favored a reinterpretation of Marx in a structuralist code, and recommended a "return" to Freud through Lacan. In effect, Althusser maintained the possibility of a synthesis between Marxism and psychoanalysis, between the sociological theory of ideology and the psychoanalytical theory of the unconscious, thanks to the notion of over-determination, that is to say a determination neither univocal nor directed to a specific end, but product of diverse determinations coming from different spheres of relative autonomy.[13] This was enough to legitimate psychoanalysis among many intellectual sectors,[14] and purge it of its suspect condition as a "bourgeois ideology" that some on the left had ascribed to it. From the beginning, Lacan´s writings were the object of interest of philosophers and literary writers that approached psychoanalysis via Sartre, first, and Althusser later. Intellectual circles not only focused on Lacan's ideas. Other thinkers involved in French structuralism began to circulate in intellectual circles, with works translated into Spanish from an early stage. In short, Masotta was one of those participants in the post-Peronist intellectual field that began in Sartre and ended in Lacan. In the psychoanalytical field, the reception and diffusion of Lacan's thought was slow and difficult, due in particular to the hegemony of the theory of Melanie Klein, which was predominant in Argentina since the 1950s. Nevertheless, this did not stop some psychoanalysts from gaining access to some of Lacan's works, including using them in the writing of scientific articles. This early diffusion was strengthened by the crisis of the APA during the 1960s and 1970s. This institution held a monopoly on the training and professional exercise from a highly professional and a-political position, consolidated during the 1950s, which had allowed it to acquire autonomy from state control, universities and medical institutions. But in the 1960s a transformation caused by the presence of a larger number of politicized psychoanalysts began. As an expression of this trend, the group "Plataforma" questioned the closed professional training of the associations, and its bourgeois ideology which pretended to be neutral. The demands were linked to the organization of the psychoanalytical career, the requirements for entry to the institution and the professional ideology.[15] The opinions of Pichon Rivière were identified, in large part, with those which fed the processes of the political radicalization and shift toward political thought that brought on people who participated in the groups "Plataforma" and "Documento" to resign from the APA in 1971. Even though Pichon Rivière supported this movement, he never decided to step down from the APA. His well-known position was considered an expression of a kind of psychoanalysis "open" to the intellectual and political world; in this way he represented a psychoanalytical and intellectual perspective that was politically acceptable for those in opposition to the APA.

Although "Plataforma" and "Documento" did not have the explicit aim to develop theoretical changes and a few of its members had studied Lacan's works, what is outstanding is that they publicly set up the image of the APA as an institution that reproduced an orthodox version of psychoanalysis. Members of "Plataforma" and "Documento" maintained that the APA wished to continue monopolizing the teaching and practice of a "true" psychoanalysis which lacked foundations. In fact, Lacanian theory could be considered legitimate by the fact that it also shared an opposition to the official psychoanalysis because Lacan had already been expelled from the IPA in 1953. On this basis, it was much more feasible that other psychoanalytical currents could begin to circulate; and what is more important that the training could be offered by other institutions, groups and individuals (like Masotta). From Lacan's point of view, even psychoanalysts could practice the profession without completing the APA's requirements (as it shortly occurred later with psychologists and with other psychoanalysts coming from the literary field, for example). This was the place that the psychologists had, new professional actors who debated the monopoly over psychoanalytical training and practice that the APA as an official institution, reserved only for doctors.

Precisely, the appearance in the psychiatric and psychoanalytical space of the psychologists graduating from the official undergraduate degree programs in Psychology at the national universities in the second half of the 1950s established a dynamic factor which contributed to the diffusion and expansion of Lacanism. From the early 1960s, the professional identity of psychoanalysis was not only converted into an aspiration but into an effective identity of psychologists as well. This caused a demand for psychoanalytical training which nonetheless, the APA could not satisfy due to the fact that it only admitted doctors among its candidates.[16] Given this, physicians had little access to the APA because there were few training analysts. Many chose to attend private study groups, sometimes led by members of the APA. At the same time different alternative training spaces appeared which in the beginning could not award the title of psychoanalysts to students. Doctors and especially psychologists began to work in the psychiatric services in general hospitals for free and voluntarily, they understood that those spaces were good places for training. In hospitals, the psychologists continued to be subordinate to doctors, a situation which was tolerated and viewed as "natural" in the early 1960s, due to the difference in experience, gender and age among doctors and psychologists; but by the end of the 1960s, and with a different political atmosphere, the conditions returned to be unbearable, so the psychologists questioned the APA's authority calling it authoritarian, reactionary and dogmatic.[17]

Questioning the sovereignty of the IPA, its pretension to maintain the heritage of training and control over the professional practice, Lacan and his theories offered to the psychologists the possibility of legitimizing psychoanalytical practice and identity not managed by the APA. Precisely, in 1969 when Masotta gave the first sign of a public confrontation with that institution in a debate with the president of the APA, he chose the first issue of the journal of the brand-new Asociación de Psicólogos de Buenos Aires, which hoped to bring together graduates from the degree program recently created at the UBA;[18] thus, the confrontation was not limited to a theoretical sphere, but the dispute also included the struggles among psychologists to seize the APA's monopoly over the profession.

In sum, describing the changes in the intellectual and psychoanalytical field and referring to the origin stories (the 1960s), it is possible to understand the passage of Masotta to Lacanism in the wider context of the transformations that took many participants in the intellectual field to the passage from Sartre's existentialism, to Marxism and to French structuralism. Also, it is possible to understand more accurately the reception of Lacan's works as part of intellectual and theoretical mutations, but also in relation to the loss of authority of the APA and the appearance of a new actor on the scene that demanded psychoanalytical training, the psychologist. Nevertheless, what this does not explain is why the meeting between Masotta and Pichon Rivière was transformed into the origin of Argentine Lacanism.

The invention of a genealogy of Lacanism in Argentina

The different versions of the origin story of Lacanism in Argentina refer to the intellectual and psychoanalytical fields of the 1960s; but in a similar way to the mythical narrations of the tribal societies, this reference stressed some contrasts and differences, and not others.[19] Through the selection of events, people and relationships, the story considers, firstly, the place of Masotta as a diffuser of Lacan. Really, Masotta had not been the only specialist and promoter of Lacan in Argentina.[20] But the story was committed to establishing fatherhood to the Lacanian genealogical origin in Argentina, and the story was able to convert Masotta in the exclusive decoder and publicist of the Lacanian message. Masotta's relationship to Pichon Rivière is what gave Masotta an exceptional place in the story. This relationship appeared with a double character. On one side, the generosity and affection that caused Pichon Rivière to give Masotta the manuscripts of Lacan, suggests that between them existed a close affinity: friendship, cordiality, respect and/or admiration. One of the versions affirms that "Pichon Rivière is evoked by Masotta as a teacher", insinuating the possibility of thinking of the relationship of generations, that is to say Masotta as a legitimate descendant of Pichon Rivière. Moreover, versions of the story maintain that Pichon Rivière gave Masotta the manuscripts because the latter understood them, or because he was especially interested in them. Therefore, while on one side the relationship stressed the harmony and solidarity between both characters, on the other it underlined opposition. On the one hand, the transfer of the manuscripts from the bookshelves of Pichon Rivière into the hands of Masotta represents continuity with the past. On the other hand, the reasoning behind the act of exchange rests in the incomprehension or disinterest of Pichon Rivière, that is how the comprehensive skill and interest of Masotta, expressed a discontinuity or break with the past. In other words: the story simultaneously affirms a dubious and contradictory relationship between the present (Masotta) and the past (Pichon Rivière). The contradiction between continuity and break with the past could be interpreted as a consequence of the argument of the story, which sustains that the origin of Lacanism in Argentina comes not only from the sole effort of Masotta, but due to his relationship with Pichon Rivière. But, why was the focus placed on the relationship between both men, and not just on Masotta?

Certainly, the story fulfilled a devoted function for the incipient Lacanian movement in Argentina until 1978, the year in which Germán García published *La entrada del psicoanálisis en la Argentina*. Born in 1944, in Junín, Province of Buenos Aires and son of a metallurgical worker, García was not a psychologist, but neither had he taken on university studies. His first calling had been literature, publishing his novel *Nanina* (prohibited by the military regime) in 1968, *Cancha Rayada* in 1970 and *La vía regia* in 1975. Between 1968 and 1970 he met Masotta, interested in his courses on structuralist linguistics and shortly after he joined the study groups on Freud and Lacan. Besides, he began to analyze with the psychologist Ricardo Malfé, coming from a Pichon Rivière perspective.[21] Coming back to what I have mentioned about the book *La entrada...*, García affirmed there that thanks to the meeting, Pichon Rivière authorized Masotta to talk about Jacques Lacan. In this act of exchange, Masotta was invested with a power which confirmed him as the unique transmitter of the Lacanian message. As a result, a new time had started, García said. The text assured that the meeting updated another exchange in 1885 which occurred in the Salpêtrière hospital, in Paris, between the French neurologist Jean-Martin Charcot and Freud, when the former had passed on to Freud a theory on the hysteria that would not be psychoanalytical yet, but opened the possibility for the growth of psychoanalysis. Similarly, Pichon Rivière carried out a theoretical renovation, questioning the APA's Kleinism. French surrealism and authors such as Georges Bataille and Conde de Lautréamont nourished his psychoanalytical viewpoint. According to García, Masotta could share the critical attitude for Kleinism and the APA, and the love for French poetry and literature. However, Masotta did not share the same theoretical perspective; hence Masotta agreed with Pichon Rivière "tactically", because he did not adhere to a theory that proposed a search for a social supplement of individual psychoanalysis, when Freud had already written that the unconsciousness was neither individual nor collective. He could not subscribe to a perspective like that of Pichon Rivière, which he considered a way of manipulating the fellow man.[22]

Putting emphasis on the position of Pichon Rivière, the story proposed a continuity with a local psychoanalytical fatherhood. In this way, it suggested that it was possible to think about the intellectual descendants of Pichon Rivière as an independent genealogy in regard to the APA, now that Pichon Rivière demanded a theoretical change (from Kleinism to social psychology), professional change (greater dedication to public health and the practice in hospices, away from the private medical centers), institutional change (a greater opening, represented in the acceptance of the training of psychoanalysis coming from non-medical degrees) and political-intellectual change (a definitive relationship with the politicized field of the intellectual left, to the extent that the system of capitalist political oppression was made clear or was reflected in the psychic disturbances and in the theories by means of which they were treated). The identification to aspects of the tradition of Pichon Rivière assured that Lacanism would continue with this tradition as much for the non-conformist sector of the official psychoanalysis, as with the critical intellectual field during the second half of the 1960s.

Pichon Rivière was presented as an ambiguous character then, "extraterritorial", a condition which permitted Masotta and his descendants to establish successfully the genealogical relationship with the psychoanalytical local past. Symmetrically, Masotta took an "eccentric place in relation to the centers of Argentine psychoanalysis", a condition that therefore allowed him to understand the message of Lacan (incomprehensible for those who took on the position as official psychoanalysts), and that converted him into the optimal candidate for bringing about the "return to Freud" as a purification, and beginning a new genealogy that would break with the official psychoanalysis, and at the same time restoring a true line of Freudian descendants.[23]

The story also allowed for the explanation of how psychologists could be converted into psychoanalysts with full rights, without the necessity of having to go to the APA or studying for a medical degree; that is how "Pichon Rivière became the hero of the excluded ones and founded a school that tried to wake up the psychologists", and for that motive "it is Masotta, then the one who took up again the dream (...) of Pichon Rivière: eliminating medical profession as a precondition in the training of psychoanalysts".[24] Given that the story declared that the condition of the psychoanalyst was not chained to the APA or to medical degrees, the place of Masotta in the psychoanalytical field could be considered absolutely legitimate. Accordingly, Masotta's successors could name themselves as "psychoanalysts", despite the fact that most of them were neither psychologists nor did they hold university degrees.

But the story was also related to a specific situation in the Argentine psychoanalytical field. García included the origin story in his 1978 book and in another book edited in Barcelona in 1980. In 1978, García was still residing in Argentina, but Masotta had already moved to Spain. In 1980, García left Argentina and moved to Barcelona, being in charge of institutional tasks since the death of Masotta in 1979. During this period, the situation for the Argentine psychoanalytical group managed and led by Masotta was difficult. In June 1974, just three days before the death of president Perón, in the middle of an atmosphere characterized by violence brought on by guerrilla groups on one side, and the ultra-right Argentine Anticommunist Alliance (the Triple A), on the other, Masotta along with a group of disciples decided to found the Escuela Freudiana de Buenos Aires. Masotta's decision to create the Escuela was extremely controversial, to the extreme that three of his followers voluntarily banned themselves. His critics claimed that the dramatic political situation in Argentina at that moment had not been mentioned in the foundation minutes of the Escuela. For his former followers, this was unacceptable and contradictory for an institution that proposed to be critical of reality. However, in December 1974 Masotta left Argentina for London and then went to Barcelona. Consequently, this departure was attributed to the growing threats from the Triple A; some maintained that the military coup led by Augusto Pinochet in Chile 1973 and subsequent murder of the constitutional president Salvador Allende influenced Masotta's decision; also including that he had been affected by the circumstance of a friend, who was held prisoner at the Estadio Nacional in Santiago de Chile.[25] Others, without ruling out the influence of the political climate in Argentina, maintained that Masotta wished to receive his doctorate degree in Oxford, with the ambition of spreading psychoanalysis and Lacan in "virgin" territories like Spain.[26]

Despite the distance, Masotta tried to continue to guide the school, through periodic epistolary communication.[27] Apparently, this style of supervision was not accepted by everyone; probably, Masotta's distancing, his pretensions of continuing to influence the government of the school, and the succession rights that some of its members had demanded by those closest to Masotta (or because they had given him a special place) brought on a growing discomfort. After several public letters were sent to the Escuela administration, on June 19, 1979 Masotta proposed to the group "to renovate the pact of 1974", continuing under another name, the Escuela Freudiana de la Argentina. Up until today, both schools assert that they were founded in 1974. Moreover, some maintain that there was a split, while others deny it. The reasons for the division remain unclear; some interpreters sustain that those who came from the medical field did not easily accept the kind of psychoanalysis without clinical practice that Masotta stood for;[28] others retort that doctors had decided to take over the Escuela, making the defense of a secular psychoanalysis indispensable.[29]

In short, the story consecrated Masotta as a non-medical psychoanalyst, and this is how the tale fulfills the objective of legitimizing Lacanism in Argentina. Using Masotta as a symbol, the story established that those who came from traditions outside the medical and the official psychoanalytical fields, best understood the message of Masotta and became his genuine beneficiaries.

Masotta, the diffusion of Lacan's ideas in Barcelona and the institution for his genealogical fatherhood

Masotta migrated to London with his wife in 1974, where he lived for about a year and half in the house of a friend. In London, Masotta began to give classes in the Arbours Association and in the Henderson Hospital of Surrey; which allowed him to build up relationships with some of the Spanish psychiatrists who worked in the hospitals, drawn by the antipsychiatric currents that inspired Ronald Laing and David Cooper. Some of them later formed part of the construction of the first institution created by Masotta in Spain. He was also able to interview the Galician psychiatrist José Eiras in Oxford, who invited him to lecture in Vigo in 1976. There even existed references of a frustrated visit to Bilbao in 1974.[30] In 1975, he began to travel to Barcelona, invited by the psychoanalyst Marcelo Ramírez Puig, just when the Spanish political and intellectual situation was at the point of a drastic change, given that Francisco Franco Bahamonde would die on November 20 of that same year. He also received invitations from the Catalan philosopher Eugenio Trías Sagnier, whom he had met in Buenos Aires. In 1976, he presented in Paris his brand-new Escuela Argentina before the École Freudienne, and he had an interview with Lacan; in that year he definitively settled down in Barcelona where he began to give classes in the studio of the painter and printer Josep Guinovart. During the course of that year, Masotta gave lectures at the Goethe-Institut, at the Fundació Joan Miró and in Vigo. Several of these interventions, together with written works in the early 1960s would be published by Catalan editors.[31] Masotta's classes were publicized by the *Revista de Literatura*, which intellectuals and writers who belonged to the movement of the post-Franco ideological and cultural renovation contributed to. The intellectuals from the *Revista de Literatura* also shared the objective of "the return to Freud" through the diffusion of Lacan's thought. In February 1977, Masotta founded the Biblioteca Freudiana de Barcelona, accompanied by his friends from *Revista de Literatura*[32] that maintained contact with the Escuela Freudiana de Buenos Aires (later "de la Argentina"). Under his impulse, the Biblioteca Gallega de Estudios Freudianos was founded, and he translated some of Lacan's works, which were published in Barcelona, and he wrote many prologs; he held two psychoanalytical conferences (1978 and 1979) at the Fundació Joan Miró; and he spread Lacan's thought to Madrid, Vigo and Valencia. Masotta continued his hard work until September 13, 1979, when he died of cancer.

Germán García replaced Masotta in February 1980. García was Masotta's disciple who formed part of the group that regrouped in the Escuela Freudiana de la Argentina. His first tasks were to direct the organization of the Biblioteca Freudiana after Masotta's death, but there he found a strong opposition. In a letter dated, April 30, 1980, five members resigned due to – as they stated – an institutional climate that championed defamation, lack of respect and obliteration of an open space for debate and differences. Those who resigned understood that the course that the Biblioteca had taken implied the destruction of the project created by Masotta in 1977; tacitly, the accusations were against García.[33] In the letter, Masotta was presented as the "founder" and "teacher," at the same time that "the promoter of the teachings of Lacan in a systematic and progressive way, in Spain". Apparently, those who renounced had not accepted the institution of "la passé"[34]. It is not this institutional conflict that interests us here – one more on the extensive and never ending chain – but the fact that the resigning group proclaimed Masotta as a "teacher". Masotta was revered for transmitting Lacan in Spain, and in virtue of this, the founder not only of the Biblioteca Freudiana de Barcelona, but also the Lacanian movement on the Iberian Peninsula. But for the resigning group, the original project of Masotta had been demolished. It was not Masotta's death that had put an end to his work, but the change in identity of the Biblioteca after the arrival of García. What is most important is that this first public conflict would cause a controversy in the Biblioteca, which served as a favorable backdrop for Masotta to be dubbed "the father" of the dissident group. Paradoxically, the resigning group impulsed Masotta's fatherhood while rejecting

García's pretension of taking Masotta's place. But it was García who a few years earlier had established Masotta and the Argentine group as the legitimate founders of Lacanian geneology in Argentina.

In July 1980, García published in Barcelona the book *Oscar Masotta y el psicoanálisis en castellano*.Different from *La entrada...*, the title showed the displacements of Masotta and García himself from Buenos Aires to Barcelona as active diffusers of Lacan. In this case, the transmission was not focused on a nation, but on the language shared by the Argentine and the Spanish people. The text was dedicated "to the friends of the Escuela Freudiana de la Argentina and the Biblioteca Freudiana de Barcelona, because they transmitted the name and supported the legacy"; a photo adjoining Masotta specified what name was transmitted and what legacy stood. García pointed out that this book would be the first in a series dedicated to the transmission of the oral teachings of Masotta;[35] so the text affirmed the continuity of a genealogy started in Buenos Aires, refuting the pretensions of those, in Barcelona, who could call themselves legitimate disciples.

Shortly afterward, on January 30, 1981 the Escuela de Psicoanálisis de la Biblioteca Freudiana was created. In the foundational minutes, written by García, the new authorities appealed to the person of Masotta to justify the foundation of the new institution. Masotta was implicitly presented as the founder of the Biblioteca, and as a generous giver of knowledge (was he perhaps reproducing the generosity that Pichon Rivière had for him?). The text also affirmed that "in Vigo (shortly before his death), Masotta confirmed the failure of the Biblioteca: he said that the Biblioteca Freudiana did not work because it had been transformed into a narcissist scene of students who converted themselves into 'outstanding' ones".[36] Along this, García gave legitimacy to his institutional project through invoking the person of Masotta. And Masotta conferred authority to García's institutional project, which aimed at establishing new rules for psychoanalytical training and practice.[37] Invoking Masotta´s words, García achieved his purpose: the adversaries had not been recognized by Masotta himself, and it was Masotta who put the blame on them. So, they could not be considered authentic Masotta's descendants, because of their arrogance. García declared that the creation of the Escuela de Psicoanálisis represented a project already set up by Masotta. He maintained that "The Biblioteca Freudiana de Barcelona, from its foundation and until this date, has been implied in an Escuela (if this word means a certain direction in the psychoanalytical theory). For this reason the style of practice of the Biblioteca was called Freudian since it spread that psychoanalytical direction by means of seminars, dissertations, publications and conferences. This means that from the beginning it was not limited to a library project, but that coming from the same, it expanded in different moments in the history of psychoanalysis in the country, taking an active position".[38] Evoked as a prophet, Masotta legitimized the decisions made in the present. Therefore, opposing the project of the Escuela was the equivalent of opposing Masotta. Masotta's death and the arrival of García speed up the creation of the Escuela, which was presented by its proponents as a triumph of Masotta's project; position completely different from that of its opponents, who alleged that it was its destruction. Therefore, the Escuela represented the continuation of the past for some, and the discontinuation for others.

Both, Masotta and then García were in charge of prolonging the efforts developed in Argentina to diffuse Lacan's thought,[39] but in a very different context. As it follows, the first difficulty to sort out consisted in knowing if psychoanalysis had existed in the Spanish past, and in such a case, what relation it had with the present psychoanalysis.

The dispute over the existence of psychoanalysis in Spain

We said that the psychoanalytic context that Masotta found in Spain was very different from the one he had known in Argentina. In Argentina, an association recognized by the IPA had been built thirty years earlier, there was a growing demand for training and care and a significant presence of psychoanalysis in the public health institutions treating psychiatric problems, and finally a large interest on the part of the intellectuals to use psychoanalysis as an interpretive instrument of social reality. In addition, psychoanalysis had been increasingly accepted by the Argentine urban middle class as part of their identity. In Spain, the situation was remarkably different. There existed an association that was recognized officially, but with only a few members and scarce insertion in hospitals and public neuro-psychiatric units. This situation gave place to a series of interpretations of the psychoanalytical past on the part of the new Lacanian movement in Barcelona, in which it was fluctuating between the negation of the previous existence of psychoanalysis, to his recognition.

In 1980, García affirmed that psychoanalysis in Spain during the 20th century had disappeared; even more, he asked if psychoanalysis had really existed in Spain. In his opinion, this so-called absence of psychoanalysis could be attributed to the Civil War (which he considered a "break" in the Spanish history) and the Franco´s regime; however,

García realized that these dramatic events could not explain everything. García understood that there were deeper reasons which explained the unfortunate destiny of psychoanalysis in Spain: ideological, intellectual and cultural reasons: thus, psychoanalysis found enemies not only among the defenders of "spiritualist" conceptions of men, but also among those who defended empiricism of experimental psychology.[40] Three years later, García modified his thesis, because he became aware of the fact that Lacan's ideas coming from Argentina did not find a desert in Spain: he maintained that Freudian psychoanalysis, which had arrived early in Spain, was slowly substituted by the Jungian and phenomenological trends. García put the blame of the main responsibility of the exclusion of Freud on the philosopher José Ortega y Gasset (1883–1955), who made a peculiar reception of Freud minimizing the importance of sexuality. At the same time, García credited Ortega y Gasset and his followers with a contemplative attitude in the face of the political conditions that ended in the Civil War and would stress the debacle of Freudian psychoanalysis.[41] Similarly, a member of the Escuela assured that Ortega y Gasset's objective in diffusing the Freud's works was to "Germanize Spain". However, in 1925, *Revista de Occidente* (publication founded by Ortega y Gasset in 1923) abandoned Freudian ideas to adhere to Jüng's perspective. According to this Escuela's member, Jungian ideas were much closer to a spiritualistic world view.[42]

As we will see, these interpretations of the Spanish psychoanalytical past were communicated through strong verbal fights. The Escuela's members argued with analysts of associations tied to the IPA, because these last ones asserted that they had restituted a psychoanalytical lineage in Spain in the midst of the Franco's dictatorship. Nevertheless, *everyone shared the same interpretation of the Spanish psychoanalytical past, which was seen as a frustrated or incomplete history.* The interpretation maintained that after an early receptive era, resistance to psychoanalysis had come out of the Spanish society, due as much to a supposed difficulty for the secularizing of collective, cultural and political life, as to the predominance of a positivist, biological and mechanicistic focus. While spiritualistic perspectives about human behavior could not accept a central place of sexuality, positivist outlooks considered notions such as "libido", "pleasure" or "desire" scientifically vague and, so, unacceptable.

According to other versions – and against the opinion of the Escuela – the hostile environment in Spain in which psychoanalysis was founded began to dissipate from the late 1920s, with the translation of the complete works of Freud. Ortega y Gasset entrusted Luis López Ballesteros y de Torres to translate Freud's complete works, which were published by José Ruiz-Castillo Basala; this event contributed to the diffusion of psychoanalytical theory and practice, together with a favorable intellectual climate during the Second Republic (1931–1939). Others maintained that Spain resisted psychoanalysis because of its "Jewish science" character, for challenging the established moral and for making relative all authority and social dream, reasons why psychoanalysis could not have good relations with a totalitarian regime. Finally, others said that psychoanalysis had been adopted by progressive and unprejudiced nations; on the contrary, Spain was conservative and intolerant. According to these opinions, open-mindedness and liberalization are typical values of modernity developed by Jewish communities; nevertheless, Spain did not become a modern society because of lacking an influential Jewish intellectual group.[43]

Against these interpretations, historian Thomas F. Glick has shown that Freud's works formed part of the great intellectual debates of the 1920s. Psychoanalysis was spread early on among readers who could read German; just like Albert Einstein's Theory of Relativity, psychoanalysis was largely accepted among the elite. Only a small and reactionary group opposed the psychoanalytical ideas. During the 1930s, the debates between conservatives and liberals intensified, deepening with the defeat of the Second Republic and the advent of Franco's regime after the Civil War. The government enacted a brutal repression over the defeated (liberal democrats, regional nationalists, socialists, communists, anarchists), over the linguistically and cultural autonomies of the different regions, and practiced censorship of ideas. Besides, the regime decreed explicitly the end to the post-illuminist science, in efforts to reestablish the "Catholic unity of the sciences". This implied the ideological control and proscription of scientific theories like Darwinism, Relativity and psychoanalysis. It was not until the end of the 1950s that psychoanalysis was reintroduced to Spain via the United States, France and Argentina. Glick's study is not only relevant because it proved the existence of an early diffusion of psychoanalysis in Spain, but also because it showed how psychoanalysis was reintroduced to Spain through different routes of diffusion coming from abroad. And one of them was Argentina, a country which was peripheral to the centers of the psychoanalytical world powers at the time.[44]

Before the Civil War, there existed a diffusion of psychoanalysis that did not crystallize in a movement.[45] It was assimilated by psychiatry, and many of the concepts were incorporated into the medical field, revolutionizing medical thinking and practice during the 1920s and

1930s. Glick showed how surprising it was that the most important diffusion and reception of psychoanalysis in Spain took place during the Miguel Primo de Rivera's dictatorship (1923–1930). According to Glick, the reason for this was that there existed a public discourse that favored scientific research and thinking. While the regime adhered to an obscurantism clerical character, the universities were centers for the free debate of ideas. Also during this time, Frances Tosquelles i Llauradó began to apply psychoanalysis in a psychiatric institution, the Instituto Pere-Mata de Reus (in Catalonia), where many Central European psychoanalysts who had run away from Nazi Germany were working. Also, the reception of Freud's ideas by Ortega y Gasset, commonly undervalued by the recently formed Lacanian movement, has been revalued more recently.[46]

Surprisingly, the openness to science ended under the Second Republic, before the Civil War and the rise of Franco's regime, due to the deep ideological polarization. Replicating the attacks that Darwin had received in the 19th century, Freud was viewed as determinist, materialist, liberal, pornographic, immoral and atheist. After the proscription of psychoanalysis during Franco's dictatorship, during the 1940s, an anti-Freudian perspective took hold in the psychiatric field, in agreement with the Nazi Germany psychiatry; some well-known names were those of colonel Antonio Vallejo Nájera (chief of the Military Psychiatric Services of Franco) and Juan José López Ibor.[47]

Nevertheless, at the height of the Franco's dictatorship, new attempts to start up a psychoanalytical reactivation appeared. In 1951, Margarita Steinbach, an analyst belonging to the recently re-established Deutsche Psychoanalytische Verbindung, moved to Madrid, where she worked as a training analyst until her death in 1954. At the end of the 1950s, a psychoanalytical group was constituted in Barcelona; this group derived from the Centro de Estudios Antropológicos y Humanísticos Erasmo, founded in 1947, where young doctors interested in psychiatry and unhappy with the biological focus began to meet. In 1954, the Sociedad Psicoanalítica Española had been constituted in Madrid; this institution received recognition from the Spanish government, but not the IPA. Due to the presence of Garma in Buenos Aires, many Spanish analysts chose to cross the Atlantic to continue with their training in the APA.[48] In 1957, the Sociedad Luso-Española was founded, and two years later it was recognized by the IPA as a member institution. When its Portuguese branch separated from it in 1967, the Sociedad changed its name to the Sociedad Española de Psicoanálisis (SEP). Finally, in 1971 the Madrid natives also left.[49]

In conclusion, there was a particular reception and development of psychoanalysis in Spain, despite the long Franco's dictatorship. For the new Lacanian movement, establishing close relationships with this preceding psychoanalysis meant, without a doubt, something problematic. As we will see, theorizing a psychoanalytical void before Masotta's arrival, or relating pre-Lacanian psychoanalysis to Franco's regime, were consequences of how difficult it was to think the liaison and the transition between the democratic present and the dictatorial past. Institutions and persons could be suspected of complicity with the dictatorship. Consequently, the SEP and its members had to prove that they were not accessories to Franco's tyranny. A self-called "psychoanalytical" institution related to Franco and his time could not be acceptable from a democratic point of view; and for this reason, its same psychoanalytical distinctiveness was doubtful. So, genealogical debates after Franco's death attempted to answer the question to when the "authentic" psychoanalytical movement had really begun in Spain.

The Lacanian genealogy in Spain as a break with Franco's dictatorship

Lacan visited Barcelona in 1958 and in 1972, invited on this last occasion by Ramón Sarró (1899–1993), a psychiatrist and a professor at the University of Barcelona, and former assistant of Wilhelm Reich at the Freud's Psychoanalytic Polyclinic of Vienna. It seems as if these visits did not awake great interest among Spanish psychiatrists. As the versions coming from the Lacanian field insist, the introduction of Lacan was developed up to the 1980s by an Argentine, Masotta. At the same time, the stories had insisted in presenting Masotta as the founder of a psychoanalytical movement with wide diffusion in Spain; this was very different from what had been generated since the beginning of the 20th century, and what had been constituted around the SEP in 1967. The Lacanian writings emphasized the place that Argentine analysts had in the diffusion of psychoanalysis in Spain, not only introducing Lacan but also psychoanalysis, that had been absent since the times of Ortega y Gasset.[50]

After Masotta's death, the Escuela argued on several occasions with SEP's members about who had reestablished psychoanalysis in Spain. In 1983, paraphrasing the famous introduction to Marx's *Communist Manifesto* ("A specter is haunting Europe – the specter of communism"), García published a reply of an article that appeared in the newspaper *El País*, titled "Barcelona es el centro mundial del psicoanálisis en lengua castellana". This article defended psychoanalysts from the IPA, underestimating the development of the Lacanian trend that Masotta and García led. Together with the letters from readers, the article questioned the Escuela for its lack of psychoanalysts in the terms accepted by the IPA; besides, the article and the letters affirmed that both the Escuela and García represented a totalitarian thinking, which exhibited "an obsolete third-world terminology". García answered pointing out that the expressions reflected an aversion to the Argentine people; and against the accusations of totalitarianism, he could show how Masotta and the Escuela had democratically spread Lacan's work throughout Spain.[51]

The Lacanian authors did not fail to recognize the existence of an established psychoanalysis under Franco's regime;[52] moreover, they had expected to differentiate genuine psychoanalysis from another spurious psychoanalysis, associated with the dictatorship. This led to the question of when the Spanish psychoanalytical movement began. Was it with the complete Freud's works published in Spanish at the request of Ortega y Gasset in 1923? Was it with the official recognition of the SEP as member of the IPA in 1959? Or was it with the arrival of Masotta in 1975? The SEP demanded being viewed as an institution that victoriously struggled against Franco's regime. However, the Escuela criticized this association, questioning with irony why an institution which was founded in the late 1950s had published its first journal in 1984, several years after Francos death. The Lacanian movement, from the beginning, created many journals: *Síntoma* (1981), *Tyche* (1982), *Serie Psicoanalítica* (1981), *Otium Diagonal* (1983) and *Entorno* (1983).[53]

Some recently published texts had tied emphatically to Masotta and the Lacanian movement in Spain with Franco's death and the return to democracy. Appealing to the opposition between life and death, the authors opposed "the thought of a living author" like Masotta that revived the intellectual and artistic world in Spain, to the agony of the dictatorship and his leader. Others suggest that Masotta found a favorable milieu, due to the recovery of the democratic freedoms and a bigger liberalization, which allowed a good will to receive new ideas and trends coming from abroad.[54] Nevertheless, according to Anne-Cécile Druet, the fall of the regime after Franco's death did not mean a break in a strict sense, since the theoretical currents and institutions that had been developed during the Franco's government endured the "democratic transition". Surely, the abolition of censorship changed the conditions for the reception and circulation of ideas, but it did not imply the indiscriminate reception of everything that was prohibited. Truly, the arrival of Masotta did not cause a massive cultural revolution, since psychoanalysis continued to be something marginal within Spanish literature, art and philosophy. Even, the studies about the cultural transformations after Franco's decease did not focused on psychoanalysis. But as we have seen, the Lacanian movement created after the arrival of Masotta gave meaning to the Spanish psychoanalytical past as a "break".[55] This rupture simultaneously referred to the remote Spanish psychoanalytical past (incarnated in Ortega y Gasset), but also to the recent past (personified in the SEP). In sum, Lacanian movement represented itself as a discontinuity with the psychoanalytical and political Spanish past. And Masotta represented a civilizing hero that reinserted Spain in Europe and the world, and the father of a psychoanalytical and intellectual genealogy politically acceptable.

Conclusion

The early propagation of Lacanism in Argentina and Spain is a remarkable case of transnational diffusion of psychoanalysis from a national context that was considered peripheral in the 1970s to other considered "marginal" from the point of view of the European diffusion of psychoanalysis. The transmission in Argentina and Spain was carried out by the same agents, which implied the link and conciliation of the origin stories with the instituted ways of relating the psychoanalytical national pasts. This process of diffusion started in Argentina in the 1960s and irradiated from Spain during the second half of the 1970s demanded the invention of genealogies that gave legitimacy to people and new Lacanian institutions founded in the name of psychoanalysis; and for that, it was indispensable to reinterpret the psychoanalytical pasts completely.

The famous story that situated the origin of Lacanism in Argentina as the result of the meeting between Masotta and Pichon Rivière refers, on the one hand, to the nature of the psychoanalytical and intellectual fields of the 1960s; on the other hand, to a relationship between the psychoanalytical present, represented by Lacan, and a past rooted in those admissible sectors of the local traditions. The transformations that have taken place in the intellectual and psychoanalytical fields during the 1960s make the move of Masotta to Lacanism intelligible. In the same way, the conditions of reception of Lacan in Argentina result more understandable, seen as a consequence of the intellectual and theoretical changes in general, and the legitimacy crisis of the APA and the arrival of psychologists and their pretensions of training and psychoanalytical identity. However, the origin story stresses the meeting between Pichon Rivière and Masotta because it explained the origins of Lacanism in Argentina as a continuation with part of a local psychoanalytical tradition, which maintained a relative theoretical, professional, institutional and political-intellectual autonomy from the APA. The meeting between these two characters seen as "marginal" assured the continuity, but at the same time, showed that a break with the past was indispensable. Given that the message of Pichon Rivière was destined to a "lay analyst" (Masotta), a legitimate non-medical psychoanalysis separated from the APA could be founded.

Exiling to Spain during the second half of the 1970s Masotta and then García continued with the task of diffusion of Lacanism and the creation of institutions in a very different political, intellectual and psychoanalytical context. In Spain, psychoanalysis had had a very peculiar development, with a complicated reception characterized by resistances and condemnations, and a late institutionalization in the 1950's, during the midst of Franco's regime. Establishing genealogical ties with the Spanish political, intellectual and psychoanalytical past was a difficult mission, bearing in mind that the country had been under Franco's dictatorship from 1939 until 1975. Some interpretations from the early 1980s, which attributed the inexistence of authentic psychoanalysis previous to the arrival of Masotta, were dined by the historic studies that showed the effective existence of a psychoanalytical, though limited development. After Franco's death, the accusations of complicity with the regime did not take long to emerge. In fact, these interpretations responded to the emerging questions in the Spanish society about how to interpret and evaluate the Spanish past after Franco's decease. For the Lacanian movement, the answer consisted of assuming that the emergence in the Spanish panorama represented, as in Argentina, a break with the past. But while in Argentina this rupture was thought of as the result of a time of transition in which components tied to the past existed, in Spain the fracture was thought of as a complete one. Lacanian movement in Spain aimed to establish a break with the distant Spanish psychoanalytical past of the 1920s, and the recent past in the 1960s. At the same time, these ways to interpret the past were ruptures with a political and ideological past. From this point of view, Masotta not only brought civilization to psychoanalysis (from Argentina), but also to the political, intellectual, philosophic and artistic spheres.

Consequentially, in Spain the legitimacy of the Lacanian movement assumed a decisively political meaning: a break with Franco's age. Despite the main founders and leaders were the same, the institutionalization was very different from what had happened in Argentina.

When crisis and schism broke out in the Escuela Freudiana de Buenos Aires, some members accused Masotta and his faction of not mentioning the tragic political situation of 1974 in the foundation minutes; other members maintained that Masotta and his group were "theorists", a kind of insult due to their supposed disinterest about social reality. In fact, the incipient Lacanian movement in Argentine differentiated their personal political positions from the politics of the psychoanalytical institutions. For them, Lacanism was not a political movement, as "Plataforma" and "Documento" radical groups. So, the meaning of Lacanism as a non-committed dogma can be based on the real emphasis of Lacanian movement on professional and institutional matters, similarly to the APA, its old rival. In Argentina, when the country returned to democracy in 1983, Lacanism was seen as apolitical by a sector of the psychoanalytical field. Given that Lacanism was the hegemonic theory at that time, some analysts and intellectuals stated that such a growing expansion had been possible thanks to the Lacanian connivance with the last military dictatorship (1976–1983). Paradoxically, while in Spain, during the "democratic transition", the Lacanian movement imagined its genealogy as a break with the psychoanalytical and political past, in Argentina, during the "democratic transition", the Lacanian movement was perceived as a continuity with the recent political past, the time of the dictatorship.

Notes

1. Jonathan Xavier Inda & Renato Rosaldo R., "Introduction: A World in Motion", in *The Anthropology of Globalization*, eds. Jonathan Xavier Inda & Renato Rosaldo (Oxford: Blackwell Publishing, 2002), p. 16.
2. Sarah Franklin, "Science as Culture, Cultures of Science", *Annual Review of Anthropology*, 24 (1995), pp.164–165.
3. Nathan G Hale, Jr., *Freud and the Americans: The Beginnings of Psychoanalysis in the United States, 1876–1917* (New York: Oxford University Press, 1971); Nathan G Hale, Jr., *The Rise and Crisis of Psychoanalysis in the United States: Freud and the Americans, 1917–1985* (New York: Oxford University Press, 1995); Elisabeth Roudinesco, *Jacques Lacan & Co.: A History of Psychoanalysis in France, 1925–1985* (Chicago: University Of Chicago Press, 1990); Sherry Turkle, *Psychoanalytic Politics: Jacques Lacan and Freud's French Revolution* (Basic Books, 1978).
4. Oscar Masotta, "La fenomenología de Sartre y un trabajo de Daniel Lagache", *Centro*, 13 (1959).
5. Oscar, "Jacques Lacan o el inconsciente en los fundamentos de la filosofía", *Pasado y Presente*, 9, (1965).
6. Ana Longoni, "Oscar Masotta: vanguardia y revolución en los años sesenta". *Segundo Simposio Prácticas de comunicación emergentes en la cultura digital,*

Séptimas jornadas de artes y medios digitales, Córdoba-Argentina (2005). http://www.liminar.com.ar/pdf05/longoni.pdf

7. For an anthropological definition of *genealogy*, see John Davies, "The Social Relations of the Production of History", in *History and Ethnicity*, eds. Elizabeth Tonkin, Maryon Mc Donald & Malcolm Chapman (London: Routledge, 1989); Rosana Guber & Sergio E. Visacovsky, "Controversias filiales: la imposibilidad genealógica de la antropología social de Buenos Aires", *Relaciones de la Sociedad Argentina de Antropología*, 22–23 (1997/1998), 27.

8. Roberto P. Neuburger, "A Psychoanalytic Tango. Recent developments in Psychoanalysis in Argentina", *Journal of European Psychoanalysis*, 7 (1998), http://www.psychomedia.it/jep/number7/neuburger.htm; Silvia Yabcowski, "Entrevista a Juan D. Nasio", *La noche inconsciente – Paladium* (1986); Emilia Cueto, "Entrevista a Juan David Nasio", *El Sigma.com* (2001), http://www.elsigma.com/site/detalle.asp?IdContenido=1410; German García, *La entrada del psicoanálisis en la Argentina. Obstáculos y. perspectivas* (Buenos Aires: Ediciones Altazor, 1978), p. 245.

9. Oscar Masotta, "Comentario para la Ecole Freudienne de Paris sobre la fundación de la Escuela Freudiana de Buenos Aires", en *Ensayos Lacanianos* (Barcelona: Anagrama, 1976), pp. 240–242.

10. Enrique Pichon Rivière, "Respuesta de Pichon-Rivière a un cuestionario sobre Jacques Lacan", *Actualidad Psicológica*, 12 (1975).

11. Silvia Yabcowski, "Entrevista a Juan D. Nasio"; Emilia Cueto, "Entrevista a Juan David Nasio".

12. Oscar Terán, *Nuestros años sesentas* (Buenos Aires: Puntosur, 1991), pp. 17–26.

13. Louis Althusser, "Freud et Lacan", *La Nouvelle Critique*, 161–162 (1964); "Idéologie et appareils idéologiques d'Etat", *La Pensée* 151 (1970).

14. Hugo Vezzetti, "El psicoanálisis y la cultura intelectual". *Punto de Vista*, 44 (1992).

15. Jorge Balán, *Cuéntame tu vida*, pp. 203–209; Mariano Ben Plotkin, *Freud in the Pampas. The Emergence and Development of a Psychoanalytic Culture in Argentina* (Stanford: Stanford UP, 2001), pp. 199–208.

16. In 1954, the Public Health Ministry established that only doctors were authorized to practice psychotherapy and psychoanalysis. The resolution pressured psychiatric doctors to limit the professional practice of non-doctors, who were until then accepted by the APA. Recently, in 1985, a law authorized psychologists to practice psychotherapy. See: Jorge Balán, *Cuéntame tu vida*, pp. 132–134, 163.

17. Jorge Balán, *Cuéntame tu vida*, pp. 158–159, 165.

18. Oscar Masotta, "Leer a Freud", *Revista Argentina de Psicología*, 1(1969); Emilio Rodrigué, "Leer a Rodrigué". *Revista Argentina de Psicología*, 2 (1969); Oscar Masotta, "Anotaciones para un psicoanálisis de E. Rodrigué", *Cuadernos Sigmund Freud*, 1 (1971).

19. Jonathan D. Hill, "Myth and History", in *Rethinking History and Myth. Indigenous South American Perspectives on the Past*, ed. Jonathan D. Hill (Urbana: University of Chicago Press, 1988), pp. 5–9.

20. Mariano Ben Plotkin, *Freud in the Pampas*, pp. 162, 203.

21. See: Miquel Bassols and Vicente Palomera, "Germán García. Psicoanálisis dicho de otra manera. Pre-textos". *Otium Diagonal. Relación Periódica del psicoanálisis en España*, 2/3 (1983), 11.

22. Germán García, *La entrada del psicoanálisis en la Argentina*, pp. 242–244.

23. García offered a similar interpretation of the ambiguous positions of Pichon Rivière and Masotta: "Enrique Pichon Rivière could not have embodied the moral of the APA (goodness, family and success): admirer of Lautréamont and of the symbolists, lover of the city and nightlife, out of place in the common pastoral. He ended up outside of the APA, building an adjacent field. In this adjacent field one finds someone who left behind the fundaments of the APA (Pichon Rivière) and another who acted as if psychoanalysis not only formed there (Oscar Masotta). For Pichon Rivière, the issue is concluded, but for Masotta the idea was just beginning and the name of Jacques Lacan is transmitted as a work in progress that comes from a discovery a few years earlier". Germán García, *El psicoanálisis y los debates culturales. Ejemplos argentinos* (Buenos Aires: Paidós, 2005), p. 231.

24. Germán García, *La entrada del psicoanálisis en la Argentina*, pp. 244, 242.

25. Germán García, "Oscar Masotta (1930–1979)", in *Oscar Masotta. El revés de la trama*, ed. Marcelo Izaguirre (Buenos Aires: Atuel, 1999), pp. 338–339.

26. Nicolás Peyceré, "Líneas sobre Oscar Masotta", in *Oscar Masotta*, ed. Marcelo Izaguirre, (Buenos Aires: Atuel, 1999), p. 237.

27. Germán García, *Oscar Masotta. Los ecos de un nombre* (Buenos Aires: Atuel, 1992), p. 44.

28. Marcelo Izaguirre, "Entrevista a Norberto Ferreira", in *Oscar Masotta*, ed. Marcelo Izaguirre, (Buenos Aires: Atuel, 1999), p. 200.

29. Alejandro Sosa Días, "La transferencia del psicoanálisis a la Argentina", in *Oscar Masotta*, ed. Marcelo Izaguirre, (Buenos Aires: Atuel, 1999), p. 244.

30. Anne-Cécile Druet, "La psychanalyse dans l'espagne post-franquiste (1975–1985)", (PhD diss., Université Paris IV, 2006), p. 204.

31. Oscar Masotta, *Ensayos lacanianos* (Barcelona: Editorial Anagrama, 1976); *Lecciones de introducción al psicoanálisis* (Barcelona: Granica, 1977).

32. These moved away from the new institution shortly after, disgusted with Masotta's negotiating attitude and they would continue with the propagation of Lacan through the literary journal, *Diwan*.

33. Rosa María Calvet i Romani, "Oscar Masotta en Barcelona", in *Oscar Masotta*, ed. Marcelo Izaguirre, (Buenos Aires: Atuel, 1999), p. 118.

34. In his text, *Proposition of October 9, 1967, on the Psychoanalyst of the School* Lacan proposed a procedure that he called "la passe", which appointed the transit between the resulting experience of personal analysis and the condition of analyst. It consists in the transmission of the experience of a personal analysis ("testimony") before oaths specifically designed by the institutions, who would evaluate if the testimony corresponded to an "objective of analysis", through which the denomination of analysis of the school was given. The vagueness that is defined in this passage, the darkness of the writing and the extraordinary similarities with the mystic initiation rites prevent greater precision, and explain to a large degree the controversies and conflicts that he had stressed.

35. Germán L. García, *Oscar Masotta y el psicoanálisis del castellano*, p. 11.

36. Asociación de Psicoanálisis Biblioteca Freudiana, *Documentos*, p. 17–22.

37. It was in Spain where García, who came from literature, began to practice psychoanalysis. See: Emilia Cueto, "Entrevista a Germán García", 15

de diciembre del 2001, *El Sigma*, http://www.elsigma.com/site/detalle.asp? IdContenido=1660

38. Asociación de Psicoanálisis Biblioteca Freudiana, *Documentos*, p. 18.

39. Even if there was a coincidence about the important role that Masotta had in the transmission of Lacan in Spain, other diffuser agents tend to be mentioned, and also Argentines. This is the case of Jorge Aleman, who developed his primordial action in Spain. He arrived in 1976, and in immediate contact with Masotta. Aleman (according to his own memoirs) found Madrid a devastating panorama, without dialogues not even between the Argentines (although they had not met that many times) or among the Spanish, more inclined toward the Anglo-Saxon and German traditions, or held up in the disputes over the psychiatric reform. Same as Masotta in Barcelona, Aleman found dialogues in the midst of the artistic, literary and philosophic atmosphere in Madrid, outside of the academic environments, in meetings that synthesized the study groups of Buenos Aires and the Madrid meetings in Cafés. From these meetings originated the first Lacanian group, "Serie Psiconalítica" (which an Argentine, Sergio Larriera joined, after his 1979 arrival), which edited a homonym journal. The work of Aleman gave the standing ground for the later group "Analytica" to form, organized by José Simonovich, and the "Ateneo Freudiano de Madrid", founded by Miriam Chorney and Gustavo Dessal, editors of the journal *El Criticón*. These three groups ended up founding the Freudian Camp, the current led by Jacques Alain-Miller mostly in Spain. See: Jorge Alemán, "Argentinos en Madrid", *El Murciélago*, No. 4 (1991); Marina Averbach y Luis Teszkiewicz, "Psicoanalistas argentinos en la salud mental española", in *La psiquiatría española en la transición* (Madrid: Sociedad Europea de Historia y Filosofía de la Psiquiatría. Extra Ediciones, 2001) http://www.persona-psi.com/wp/?p=21

40. Germán García, *Oscar Masotta y el psicoanálisis del castellano*, p. 129.

41. Germán García, "Para una historia del psicoanálisis en España", *Otium Diagonal*, 2/3 (1983), 10.

42. Miquel Bassols, "Desde Buenos Aires", *Otium Diagonal*, 2/3 (1983), 44–45.

43. Marina Averbach y Luis Teszkiewicz, "Psicoanalistas argentinos en la salud mental española"; Francisco Carles, Isabel Muñoz, Carmen Llor, Pedro Marset, *Psicoanálisis en España (1893–1968)* (Madrid: Asociación Española de Neuropsiquiatría, Colección Estudios no. 26, 2000), pp. 9–10; María Luisa Muñoz, "Contribución a la historia del movimiento psicoanalítico en España: formación de la Asociación Psicoanalítica de Madrid", *Revista de Psicoanálisis de Madrid*, Asociación Psicoanalítica de Madrid, 9–10 (1989), 121.

44. Thomas F. Glick, "The Naked Science: Psychoanalysis in Spain, 1914–1948", *Comparative Studies in Society and History*, 24 (1982), 534–535.

45. With the exile of the Bilbao native Ángel Garma (1904–1993) the possibility of counting on an official institution affiliated to the IPA was frustrated, due to the Civil War. Analyzed by Theodor Reik in the Berliner Psychoanalytisches Institut, he had an analysis in control with Karen Horney, Otto Fenichel, and Jeno Harnik. Member of the Deutsche Psychoanalytische Gesellschaft since 1932, he returned to Madrid in 1931, where he worked as an analyst and trainer, and he published and participated as a member of the Neuropsychiatry Association and the League of Mental Health. In 1936, before

the outbreak of the Civil War, he left the country, leaving for Paris and then permanently to Buenos Aires where he was one of the founders of the APA.
46. Francisco Carles *et al.*, *Psicoanálisis en España*, pp. 231–232; Helio Carpintero y María Vicente Mestre, *Freud en España. Un capítulo de la Historia de las ideas de España* (Valencia: Promolibro, 1984); Vicent Bermejo Frígola, "Ortega, Freud, el psicoanálisis y la interpretación de los sueños", *Revista de historia de la psicología*, 21 (2000), pp. 631–658; Christian Delacampagne, "La psychanalyse dans la péninsule Ibérique", in *Histoire de la psychanalyse*, ed. Roland Jaccard (Paris: Hachette, 1982), pp. 383–394; Thomas F. Glick, "The Naked Science", pp. 535, 566; Antonio Sanchez-Barranco Ruiz, "Ortega y Gasset, la psicología y el psicoanálisis, *Revista de Historia de la Psicología*, 16 (1995), 255–261; Elisabeth Roudinesco & Michel Plon, *Diccionario de Psicoanálisis* (Buenos Aires: Paidós, 1998). García maintained that despite Ortega y Gasset had been displeased by the attention Freud had given to sexuality as a way to interpret dreams, he thought that it was useful as a pedagogical instrument, because it allowed changing certain atavistic traces of the Spanish. See: Germán García, *Oscar Masotta y el psicoanálisis del castellano*, pp. 131–132.
47. Thomas F. Glick, "The Naked Science", pp. 567–68, 571.
48. Francisco Carles *et al.*, *Psicoanálisis en España*, pp. 238–252.
49. There were also parallel and independent developments, as the el Instituto-Clínica Peña Retama de Hoyo de Manzanares (in the Sierra of Madrid), directed by Jerónimo Molina Nuñez, where the first therapeutic community was organized in 1962, the training base on the teachings of E. From, K. Horney, S. Ferenczi and W.R.D. Fairnbain, among others took shape.
50. Miquel Bassols, *Otium Diagonal*, pp. 44–45.
51. Germán L. García, "Un fantasma recorre España. La Asociación Escuela de Psicoanálisis de Barcelona", *Otium Diagonal*, 1 (1983).
52. Manuel Pérez Sánchez, "Inicis del moviment psicoanalític a Barcelona", *Revista Catalana de Psicoanàlisi*, 1, Institut de Psicoanàlisi de Barcelona (1984).
53. Elvira Guilanyà, "Datos para una historia", *Otium Diagonal*, 7 (1984), 88–89.
54. Marina Averbach y Luis Teszkiewicz, "Psicoanalistas argentinos en la salud mental española"; Clotilde Pascual, "La llegada del psicoanálisis a Cataluña. Los inicios y las formas del movimiento lacaniano", *Norte de Salud Mental*, 26 (2006), http://www.ome-aen.org/NORTE/26/NORTE_26_110_75-81.pdf, 78; Miquel Bassols, "Oscar Masotta, un lugar de enunciación", Fundación Descartes http://www.descartes.org.ar/masotta-bassols.htm.
55. Anne-Cécile Druet, "La psychanalyse dans l'espagne post-franquiste (1975–1985)", pp. 280–282, 397; José B. Monleón, "El largo camino de la transición", in *Del franquismo a la posmodernidad. Cultura española 1975–1990*, ed. José B. Monleón (Madrid: Akal, 1995); Samuel Amell & Salvador García Castañeda, ed., *La cultura española en el posfranquismo. Diez años de cine, cultura y literatura (1975–1985)* (Madrid; Playor, 1988); Teresa Vilarós, *El mono del desencanto. Una crítica cultural de la transición española (1973–1993)*, (Madrid,: Siglo XXI, 1998).

Index

The locators in bold refer to section heads.